MW00475079

Mathematical Discourse
Let the Kids Talk!

Critique

Justify

Question

Author
Barbara Blanke, Ph.D.

Foreword
Steve Leinwand

Publishing Credits

Corinne Burton, M.A.Ed., *Publisher;* Conni Medina, M.A.Ed., *Managing Editor;*
Diana Kenney, M.A.Ed., NBCT, *Content Director;* Veronique Bos, *Creative Director;*
Robin Erickson, *Art Director;* Marissa Dunham, M.A., *Assistant Editor;*
Lee Aucoin, *Sr. Graphic Designer*

Image Credits

p.105 (top let) Courtesy of Jessic Djruic; p.105 (all others) Courtesy of Kimberly Kelly; p.117
Courtesy of Dr. Craig Froehle; p.134 (all) Courtesy of the Math Learning Center; all other images
from iStock and/or Shutterstock.

Standards

© Copyright 2010. National Governors Association Center for Best Practices and Council of Chief
State School Officers. All rights reserved.
© Copyright 2007–2017. Texas Education Association (TEA). All rights reserved.

Shell Education

A Division of Teacher Created Materials
5301 Oceanus Drive
Huntington Beach, CA 92649-1030
http://www.tcmpub.com/shell-education
ISBN 978-1-4258-1768-8
© 2018 Shell Educational Publishing, Inc.

The classroom teacher may reproduce copies of materials in this book for classroom use only. The
reproduction of any part for an entire school or school system is strictly prohibited. No part of this
publication may be transmitted, stored, or recorded in any form without written permission from
the publisher.

Weblinks and URL addresses included in this book are public domain and may be subject to
changes or alterations of content after publication by Shell Education. Shell Education does not
take responsibility for the accuracy or future relevance and appropriateness of any web links or
URL addresses included in this book after publication. Please contact us if you come across any
inappropriate or inaccurate web links and URL addresses and we will correct them in future
printings.

Table of Contents

Appendix

Foreword

There is much we know about how to significantly improve the teaching and learning of mathematics. We know that when students are productively engaged, learning is an almost automatic byproduct. We also know that student engagement emerges most effectively from our questions and queries which open doors to student discourse. That is, it is student talk—to their classmates, with their teachers, in small groups—that is a critical component of effective teaching and learning. However, as with all aspects of effective teaching, maximizing the quality and quantity of student discourse takes practice, time, listening, trust, and a commitment to our students. That's where this readable, insightful, and helpful book comes in to play.

Effective and productive discourse does not emerge from a bunch of random questions posed to our students, nor from mindsets that teaching is primarily a process of telling, showing, and practicing. Rather, the higher levels of student learning that *do* emerge from powerful discourse within classroom communities of engaged learners requires intentionality with our tasks and questions, careful listening, and focused mindsets about empowering students. Once again, this is much more easily said than done. But how we implement these and other critical aspects of teaching are thoughtfully described and illustrated in this book.

Building a discourse-rich classroom often starts with Math Talks or number talks. Such Math Talks are safe and relatively easy ways to create a classroom culture of discourse and Chapter 4 can easily stand alone as a valuable summary of the what, why, and how of using Math Talks. Similarly, Chapter 6, where one finds 20 lesson plans to launch the school year serves as a great summary—call it a mini-math-methods course—of how to institute a range of practices and strategies that characterize effective mathematics classrooms. It too can serve as a self-contained initiator of discussions and action planning as part of grade-level teams, PLCs, and faculty meetings.

As noted in Chapter 5, "equitable participation…does not happen automatically. Teachers need to be willing to let go of the expert/evaluator role to give students opportunities to take risks and share what they think. This can be difficult for many teachers" (118). When we are honest with ourselves, we acknowledge how true this

Foreword *(cont.)*

is and what a challenge it can be to "embolden students into a meaningful discourse community" (18). Throughout *Mathematical Discourse: Let the Kids Talk!*, Barbara Blanke gives us a road map for meeting this challenge and for significantly enhancing the quality and impact of student discourse in our mathematics classrooms.

The best teachers I know and observe understand that their greatest challenge is to be learners themselves, deep thinkers and reflectors, and risk-takers. The chapters that follow provide a wide range of guidance to all teachers who aspire to meet this challenge.

—Steve Leinwand
Principal Research Analyst
American Institutes for Research

Acknowledgments

This book has been a longtime dream that began with the children I had the privilege to teach and learn from throughout my years as a K–12 teacher. When I first saw my elementary students blossom in a discourse-rich environment, I began the journey to make that environment a reality in my mathematics classroom! I was very fortunate to learn from several special people. My insights in student-centered teaching began with Mrs. Donna Burk, master teacher, author, and teacher-of-teachers extraordinaire. She led me to the path of teaching other teachers in the field of mathematics and introduced me to the Math Learning Center professional development team.

Through the incredible minds of Dr. Gene Maier, Dr. Albert Bennett Jr., Allyn Fisher, Linda Foreman, Jill Board, Trudy Mitchell, Debby Head, Libby Pollett, Michael J. Arcidiacano, Mike Wong, Pat Wong, and Kathy Pfaendler, my journey to learn more about teaching math began. The late Dr. John Van de Walle joined this elite group by becoming an email buddy who supported my first writing experiences and provoked me to think deeper about the mathematics I taught. I give gratitude to Steve Leinwand who always modeled the voice I wanted to develop throughout my career as a math educator; always talking about sensible mathematics. I am a believer that I must deliver a message to provoke action, and you have been my inspiration for many years. A huge influence in my career has been the research work of Virginia Bastable, Deborah Schifter, and Susan Jo Russell, who made me think deeply about how computational fluency influences student algebraic reasoning. All of these educators pressed me to make sense of the math and helped me to develop an appreciation of the beauty that lies in the mathematics we teach. Without them, my journey would have never begun.

I thank my colleagues and friends who taught me to persevere and enjoy the process of the journey; especially, Jill Laursen, Kimberly Kelly, Jessica Djuric, TJ Jemison, Dr. Susan McBride, Dr. Boyce Heidenreich, Amy Shields, Pia Hansen, Martha Ruttle, and Dr. Barb Bamford. The passion of these mentors for teaching is evident in the students' lives they've touched. Together, I believe we have altered the belief systems and mindsets of teachers and students relative to teaching and learning mathematics. I give special thanks to Dr. karen Higgins, my doctoral advisor, mentor, and friend who convinced me

Acknowledgments *(cont.)*

that researching discourse was important and believed in me as a teacher-educator and researcher. I also give a special thank-you to Cary Cermak-Rudolf who opened the door to her classroom every day for four months so that I could deeply understand and see mathematical discourse in action! I am forever indebted to you! And to all the teachers and students who have continually opened their classroom doors and invited me in so that I could continue to learn with and from them, you are all amazing teachers and the children's lives you touch have been forever changed!

I have special appreciation and love for my parents, Paul and Eleanor Creger, who raised a lifelong learner from the beginning and taught me to embrace learning and do everything with a passion for life. Thank you to my best friends, Susan Roberts and Dr. Bonnie Markoff, for reading, editing, taking care of my dogs, and being genuinely excited for me throughout this writing process. You are my family! And of course, having the time to write and complete this book was only possible because of my husband, Dan, for never questioning and always supporting my love of lifelong learning! And thank you to Ellie and Kala, our golden retrievers, who missed many a morning walk while I was at the computer but never left my side.

Finally, I would like to express my gratitude to Sara Johnson and Diana Kenney for persevering and convincing me that I have something important to share with other teachers. I also have deep respect and gratitude for my amazing editor, Marissa Dunham, and the creative staff at Shell Education, who have polished my ideas and fueled my passion for writing. You have provided the platform for me to deepen my thinking and share my thoughts with educators everywhere.

Introduction

Each time one prematurely teaches a child something he could have discovered himself, that child is kept from inventing it and consequently from understanding it completely.

—Jean Piaget 1970a, 715

My goal for writing *Mathematical Discourse: Let the Kids Talk!* is to support teachers choosing to shift their mindsets about how they were taught math and were taught to teach math and to encourage them to begin using discourse to help their students understand and make sense of mathematics. So many of us were taught to explain, or show and tell, students what they needed in math, and we just crossed our fingers and hoped that they would learn it. We taught math and then assessed our students to see what they could and could not do, and we proceeded to teach them what we thought they needed to strengthen the desired behavior of coming up with a correct solution. We were encouraged to offer "hands-on" learning to induce "correct" understandings. Yet, many of our students fell behind and did not learn what they were supposed to master, and we moved on to make sure we covered the skills. This book's purpose is to support any educator who desires to shift his or her practice by becoming a reflective practitioner and promote discourse opportunities as a way to develop a deeper understanding of mathematics for all students. Inventing and understanding are critical to the development of a positive growth mindset. When we, as teachers, learn to foster disequilibrium, puzzlement, discourse, reflection, and argumentation, we are encouraging students to reason about mathematical ideas. We are using talk as a means to help students to think on their own. They no longer need the teacher to explain, or show and tell.

As teachers, we have to recognize what these necessary shifts in our mindsets are that will enable us to transform our classrooms into mathematical learning communities where students seek to deeply understand mathematics. The Common Core State Standards of Mathematical Practice and state process standards guide us to make certain all students persevere in solving mathematical problems with tenacity and confidence.

We use tools strategically, explore mathematical structures using visual models, attend to precision, and pose viable arguments to critique ideas to nurture student academic growth. Creating a classroom environment conducive for students to generate a depth of understanding of mathematics—rather than being told how to get to the answer—is critical. We, as teachers, have a new role: one of facilitator and mentor, rather than explainer. We want to teach our students to do more than a share-and-tell, we want them to talk about and justify their ideas, build mathematical arguments, and defend their thinking. The questions we should be asking are: How do we support this kind of classroom on a daily basis? Where should we begin? How do we learn to *Let the Kids Talk!*?

My purpose in writing this book on mathematical discourse is to empower teachers everywhere to believe in and nurture their students' ideas, questions, and arguments as budding mathematicians to become autonomous learners in their K–12 classrooms. It was in my own classroom that I experienced the influence of mathematical discourse. It was there that I learned with and from my students rather than delivering knowledge to them. My students taught me to slow down and listen to their ideas. It was my students who nurtured and challenged me to embrace my role as a co-learner and to ask, "Is there another way to support _ALL_ of my students as they make sense of mathematics and develop the habits of mind of a mathematician?"

This book offers a brief history of mathematical discourse and takes you through specific strategies, methods, instructional approaches, resources, and activities to use with your students as they learn to make sense of mathematics on a daily basis.

The chapters include the well-researched and studied effects of discourse over the years, as well as various strategies, routines, teacher moves, and questioning techniques that will support you in transforming your classroom into a discourse-rich environment. The following chapters are discussed:

- Chapter 1, *What Is Mathematical Discourse?*, shares a brief history of mathematical discourse, some patterns that continue to influence how we teach mathematics, and why it is important to consider shifting our mindsets and teaching practices to include discourse in today's math classrooms. We explore how this knowledge can assist you in promoting mathematical conversations with your students.

- Chapter 2, *Discourse and Mathematical Practice and Process Standards,* highlights how sharing ideas, asking questions, and reasoning about mathematical ideas support the habits of mind that <u>ALL</u> students need to develop to meet mathematical practice and process standards. It provides various and specific strategies for you to build a discourse-rich community of learners.

- Chapter 3, *Teacher Moves That Promote Effective Student Discourse,* reminds us of the thousands of teacher decisions and pedagogical moves that we make daily in our math classrooms and how we can learn to reflect and act upon them to promote productive, healthy risk-taking, while building a discourse-rich community of mathematical learners.

- Chapter 4, *How Math Talks Promote Discourse: Arguments, Ideas, and Questions,* explores what specific teacher talk moves can be implemented during Math Talks, a daily routine, which offers equitable opportunities for all students while engaging in mathematical discourse. It answers the question: How do we begin to get our students to talk, justify, critique, and challenge each other about their mathematical thinking?

- Chapter 5, *Equity and Engagement,* continues the discussion about how our teacher beliefs and knowledge of our students impact student engagement with cognitively demanding tasks in an inclusive and safe classroom environment. We look at the strategies that will help you turn routine math problems into non-routine problems through the use of mathematical discourse; thereby, providing equity for all students.

- Chapter 6, *Getting Kids Ready to Talk! The First 20 Days of Discourse,* provides a day-by-day guide that answers the questions: How do I get my students to talk mathematically in my classroom? Where do I start? How do I fit this in with all the content that I must cover? This day-by-day guide will show both elementary and secondary teachers how to build community and establish necessary norms to create a discourse-rich environment where students take risks, justify their strategies, and question and clarify mathematical ideas all while deepening their understanding of mathematics.

As teachers, we all have a strong desire to meet our students' needs, and we can begin to do that by believing mathematics can be accessible for all through the use of discourse. When we choose to learn together, we engage in discourse and communicate our thoughts and feelings about what it means to be effective teachers and learners of mathematics. It is a great privilege for me to use this book as a vehicle to continue the conversation and collaborate with teachers everywhere to deepen our collective thinking and learning, to create a fascination about what others think and do, and to begin to think about how other people's thoughts might influence our own ideas and understanding. It can be joyful and enlightening to ask questions, hear heartfelt "Ahas," and see frustrating roadblocks give way to ideas that have been there all along. My dream is that this book is a stepping-stone for teachers to begin the conversation and take the first steps toward building equitable learning environments where everyone, students and teachers alike, develop the habits of mind to blossom into lifelong learners…as mathematicians.

Resources

The paperfolding lesson resource and student Math Talk reflection pages in this book are available as digital PDFs online. A complete list of the available documents is provided on page 182. To access the Digital Resources, go to: www.tcmpub. com/download-files. Enter this code: 84715886. Follow the on-screen directions.

Chapter 1

What Is Mathematical Discourse?

What does it mean to do mathematics, and why do so many of our students dislike learning it in the classroom? This is a question that pops up more often than it should. There is an uneasy resistance to math that boggles the educational mind and leaves us wondering why. Why such anxiety? One reason could be that math is seen by most students (and sometimes parents and guardians) as different from other subjects. Math is frequently taught as a performance subject, where getting to the correct solution is the only goal. They see the math that they do in the classroom setting as something disconnected from their world. For many, doing math in school is only that—doing. It means completing worksheets and getting answers, and rarely do students think about or desire to see the beauty of mathematics, to share ideas, to ask questions, to struggle productively, or to look for rich connections that can make math applicable to the real world. The nature of mathematics does not make it different, as many believe; instead, it is due to the widespread misconception that mathematics is a subject of rules and procedures, using strictly numbers, recalling facts quickly, and being fast to get right versus wrong answers. Teachers who believe in promoting mathematical discourse have seen students begin to change this baseline traditional view of mathematics and begin to see mathematics as a creative, visual, connected, and exciting subject. Students who engage in mathematical discourse are given this opportunity in many different ways. Ultimately, it is you, the teacher, who shapes mathematical learning for students you teach. The first step is to understand what mathematical discourse means.

Facilitating student engagement in mathematical discourse can be both revitalizing and demanding. It can be easy to get a conversation going by asking students to share their ideas or strategies to the class. Though their ideas often amaze us as educators, mathematical discourse goes beyond students performing a show-and-tell of strategies and solutions. It's more than the class merely listening and clapping and then moving on. Knowing what to do with students' ideas and teaching them how to meaningfully ask challenging and clarifying questions during discussions can be quite critical. If done well, mathematical discourse can shape teachers' and students' perceptions of themselves as mathematicians, as well as their ideas about what being a mathematician involves.

According to Deborah Ball:

> **Discourse is used to highlight the ways in which knowledge is constructed and exchanged in classrooms. Who talks? About what? In what ways? What do people write down and why? What questions are important? Whose ideas and ways of knowing are accepted and whose are not? What makes an answer right or an idea true? What kinds of evidence are encouraged or accepted? (1991, 44)**

NCTM (2000) defines mathematical discourse as both students' and teachers' ability to articulate mathematical ideas or procedures via talking, writing, asking questions, and responding to ideas. These are realized through various configurations in the classroom in Figure 1.1:

Figure 1.1 Student-Teacher Relationships in Mathematical Discourse

Definition	Explanation
Student-to-Teacher	The student primarily addresses the teacher even though the entire class or group hears the student's comments.
Student-to-Student	The student addresses another student.
Student-to-Group/Class	The student addresses a small group of students or the entire class.
Individual Reflection	The student documents his or her reflections about mathematics in writing.
Teacher-to-Student	The teacher addresses the student(s).
Teacher-to-Group/Class	The teacher addresses a small group of students or the entire class.

As we see, effective mathematical discourse is an interactive process. Students engage in various types of discourse at different cognitive levels, deepening their questions to lead to explanations and justifications that may be challenged, defended, or clarified. Students can thereby form new generalizations, initiating new conversations.

Most notably, it is the student who is in control of the conversation, and the teacher's role is that of discussion facilitator. It is these student-to-teacher questions, generated from conversations, which lead to explanations and justifications that may be challenged, clarified, and defended within the discourse classroom community. During discourse, students will often make new generalizations and conjectures that initiate a deeper level of student communication through the discussion of ideas, strategies, and thinking. These are some of the important things discourse provides: ways to scaffold mathematical thinking into the curriculum for *all* students. After all, it is the teacher who plays the crucial role in shaping discourse with signals to their students about how to value mathematical knowledge as well as ways to think about and know mathematics (Ball 1991; Blanke 2009).

It can be tricky to create and maintain discourse environments in the classroom for many teachers (Sherin 2002). Often, it's not enough to encourage students to discuss ideas and converse with each other. Teachers must also ensure discussions are mathematically productive and continually scaffold student academic growth. Asking teachers to use discourse requires teachers to develop a new sense of what it means to teach mathematics and to be effective, successful mathematics teachers (Sherin 2002; Smith 1996). At the same time, teachers must want to better understand their roles in developing discourse and make the choice to change to effectively implement mathematical discourse daily in their math classrooms. The choice to develop discourse communities can be nurtured through shared ideas and the desire to continually learn.

A Historical Perspective on Mathematical Discourse

We work in the fog of collective amnesia.

—Treisman 2016

Mathematical discourse is not a new idea! A careful study of mathematical history uncovers the fact that many ideas that have been called new are actually old, being revisited and reapplied into a broadened mindset. The examination of studies and research of the past has also contributed to current research of the effective usage of discourse in mathematics education. Unfortunately, there continues to be obstacles in implementing research-informed instructional practices, such as discourse, in our classrooms. A brief look at the history of mathematical discourse points out that mathematics education continues to be cyclical. This lens gives insights into the factors that may still be in contention today, the influences that continue to alter instruction, and the motives and conflicts that shape the present day learning of mathematics.

Studying our knowledge of the past empowers us to speak with more authority and to make better-informed decisions about our everyday practices in the classroom. History can offer a better insight into the influence of mathematical understanding and mathematical discourse on student learning and classroom teaching. Realizing this vision of classroom discourse in mathematics education has proven to be challenging to implement over the years. This supports the importance of studying how teachers develop and sustain mathematical discourse communities, and it supports the goal of developing a deep mathematical understanding for all. It is a goal of this book to offer teachers ways to support mathematical discourse in their math classrooms and encourage all students to construct viable arguments with others because learning is a collaborative endeavor that emphasizes the importance of every student's depth of understanding.

As stated in NCTM's *Principles to Actions: Ensuring Mathematical Success for All* (2014), there are dominant cultural beliefs about teaching and learning mathematics that sometimes get in the way of implementing best practices, such as mathematical discourse, in our classrooms. Developments in the mid- to late-nineteenth century have profoundly influenced our knowledge with regard to how students learn and understand mathematics. The very first American school mathematics textbook was Nicolas Pike's *New and Complete System of Arithmetic—Composed for the Use of the Citizens of the United States* (1788). Pike's teaching process was 1) state a rule, 2) give an example, and 3) have students complete a set of practice exercises. This scripted approach established a method of delivering mathematics to students that was rigid and has become deeply rooted in educational culture. Many today still live in the shadow of these ideas and stereotypes about learning arithmetic that was articulated so long ago.

The first attempt to change instruction was offered by Colburn's (1849) discovery learning, which offered the idea for teachers to postpone student practice until after students had developed a deeper conceptual understanding of mathematics. This began the movement toward the use of discourse and talking about mathematical ideas. His idea was expanded upon again by Charles Davies (1850) and Edward

Brooks (1880). Davies and Brooks discussed how student understanding and thinking could be strengthened through the act of studying and talking about mathematics to cultivate one's mental abilities (Bidwell and Clason 1970). A backlash to Colburn's ideas was quick and reminiscent of today's math clashes. The outcry became that rules were necessary and students could not be expected to invent them. New publications proclaimed that they would satisfy parents who longed for arithmetic to be taught "the good old fashioned way" with concise and plain explanations of rules. Today, this is still somehow a serious roadblock for developing productive discourse in mathematics.

Secondly, arithmetic became formalized through the creation of a logical system of definitions, principles, and theorems. This led to mathematics being taught based on quantity, which allowed mathematical and psychological theory to blend. The belief was students learned mathematics from a quantitative reality via student perception and intuition. Through these ideas, knowledge was built through reasoning and imagination. William Brownell's *Meaning Theory of Learning* (NCTM 1935) proposed the ultimate purpose of arithmetic instruction is the development of the ability to think in quantitative situations. The word *think* is used cautiously; the ability to merely perform certain operations mechanically and automatically is not enough. Students must be able to analyze real or described quantitative situations (NCTM 1935, 28).

The third development in mathematical discourse is an emphasis on using manipulative objects in early number work to better understand mathematics. Toward the end of the nineteenth century, the impact of this "new" psychology on mathematics teaching was presented by William James in his book *The Principles of Psychology* (Boring 1957). James provoked an extensive reevaluation of what mathematical content should be taught and why. He spoke to the debate about the origin of the meaning of *number* by asking: Does number exist without any sensemaking by the user, or does meaning evolve from the human mind? (Bidwell and Clason 1970).

So, within the first half century of the founding of the United States, the great school of mathematics debate was underway. Should teachers give students rules and facts to memorize? Or, should they offer problems to talk, argue, discover, and develop a deep understanding of the underlying mathematical principles?

During the twentieth century, and extending into the present day, the teaching of mathematics in American schools has continued this debate and experienced identifiable phases, each with a different emphasis (see Figure 1.2 on page 18).

Figure 1.2 Phases of Mathematics Education and Psychological Learning Theories

Phase	Main Theories and Theorists	Focus	How Achieved
Drill and practice (~1920–1930)	Connectionism or associationism (e.g., Thorndike)	Facility with computation	Rote memorization of facts and algorithms Break work into series of small steps
Meaningful arithmetic (~1930–1950s)	Gestalt Theory (e.g., Brownell, Wertheimer, van Engen, Fehr)	Understanding of arithmetic ideas and skills; Application of math to real-world problems	Emphasis on mathematical relationships Incidental learning Activity-oriented approach
New math (~1960–1970s)	Developmental Psychology, sociocultural theory (e.g., Bruner, Piaget, Dienes)	Understanding the structure of the discipline	Study of mathematical structures Spiral curriculum Discovery learning
Back to basics (~1970s)	(Return to) Connectionism	(Return to) Knowledge and skill development	(Return to) Learning facts by drill and practice
Problem solving (~1980s)	Constructivism, cognitive psychology, and sociocultural theory (Vygotsky)	Problem solving and mathematical thinking processes	(Return to) Discovery learning, learning through problem solving
Standards, assessments, and accountability (~1990s–2000)	Cognitive psychology, sociocultural theory versus renewed emphasis on experimental psychology (NCLB)	Math wars: concern of individual mathematical literacy versus administration of educational systems	NSF: student, standards-based curricula versus text preparation for state-specified expectations
National Research Council's *Adding It Up and Common Core State Standards for Mathematics* (~2000 to Present)	(Return to) Gestalt Theory (e.g., Brownell, Wertheimer, van Engen, Fehr) and Developmental psychology, sociocultural theory (e.g., Bruner, Piaget, Dienes)	Aim to end the "math wars" by summarizing best practice research on K–8 mathematical learning, defining mathematical proficiency as multidimensional	Standards emphasize skill development and student understanding, reasoning, and problem solving Standards-based assessments force emergence of curricula for depth of understanding

Adapted and extended from Lambdin and Walcott 2007

In 1922, Thorndike published *The Psychology of Arithmetic* in which he discussed the important role teachers of mathematics have in the learning process. He presented the necessary steps that mathematics instructors should implement to ensure every student learns and practices explicitly. His theory influenced mathematics education toward a behavioristic (stimulus-response) view of how to teach mathematics, describing the teacher's role as one that arranges for students to receive the right type of drill and practice during each of the steps and to provide the right amount of time. The major effect of Thorndike's theory was the segmentation of curriculum into many disjointed bits of information (DeVault and Weaver 1970). This emphasis toward teacher-directed learning, which instructed the practice of skills in small, distinct, and easily mastered units of information, was not seen as effective by some parents and educators in the 1930s and 1940s. The extreme emphasis on drill for drill's sake was questioned and initiated attempts by math educators to ensure basic skills were part of the bigger picture; namely, meaningful mathematics (Lambdin and Walcott 2007). These mathematicians and educators did not seem to rely heavily on any one learning theory. Their focus was to design materials that would help students understand the underlying structure of mathematics.

Psychologist Jerome Bruner (1960) became closely related to this new mathematics movement because he delved into the importance of teaching fundamental structures for any subject. His work emphasized two major ideas for the new math curriculum: the idea of a spiral curriculum and the idea of discovery learning. His work was supported by Piaget's (1952) developmental stages of learning theory. Many teachers of mathematics during this era supported these steps: 1) promote the discovery of mathematical concepts through manipulation of objects, 2) present concepts pictorially, and 3) introduce appropriate mathematical symbolism. They believed curriculum should act like a spiral where ideas are revisited again and again in increasingly more complex and abstract forms. Bruner also supported the idea of discovery learning. This was not a new idea. During these years, most textbook authors adopted a guided discovery approach. It was an important component of mathematical discourse.

We can think of these arguments as first steps in the shift from the poorly named "old way," or traditional way, of teaching mathematics to a more progressive way seen as teaching student-centered mathematics, which encourages students to talk, take risks, and justify their thinking. Traditional mathematics instruction focused on procedures to get to solutions. Student-centered mathematics require students to talk about math, share ideas, argue and reason, and question to develop a deeper understanding of mathematical concepts. Many parents, politicians, and even teachers had difficulty appreciating the new math phase because both mathematics and the form of instruction were viewed as foreign to what they had experienced in the past. In the early 1970s, these concerns led to recommendations that mathematics education go back-to-the-basics. This resulted in some schools' rapid rejection of the new math and a return to traditional mathematics, or the drill-and-practice phase. Traditional math, which many experienced, did not allow student talk or discourse; therefore, promoting the idea that there was one correct way to do mathematics. This increased tensions.

A decade after the back-to-basics phase, many educators thought the pendulum had, once again, swung too far. The 1980s focused on international competitiveness in commerce and the need for a general, competent workforce who could advance the nation's overall economic and technological progress (Davis, Maher, and Noddings 1990). This led to the problem-solving phase. The phase focused on the idea that all students should be able to use mathematics to solve contextual problems. Lessons in mathematics emphasized problem-solving strategies, and these were directly taught to students. By the end of the 1980s, the problem-solving movement was further refined as a distinction between teaching students about problem solving versus teaching for problem solving. It was in this phase that mathematics teachers were encouraged to form cooperative learning groups and encourage students to verbally share their thinking about the process of solving problems. During this time, too, both Piaget's (1952) and Vygotsky's (1978) theories on learning added to current constructivist researchers who stressed the importance of discourse as a means for developing mathematical understanding (Davis, Maher, and Noddings 1990). When students were pressed to develop strategies while solving high-demand tasks, they collaborated with each other both through written and oral discourse. Eventually, they saw learners who engaged in a high level of student-to-student discourse without the support of a teacher.

In 1989, NCTM's Curriculum and Evaluation Standards for School Mathematics began the final phase of our historical review. The further development of NCTM's *Principles and Standards for School Mathematics* (2000) and the No Child Left Behind Act (Britannica 2001) increased the focus on accountability and led to the continuation of spirited discussions about the best way to teach mathematics. Needless to say, opinions vary.

The emphasis on standards-based learning asked educators to reconsider the influence of proficiency on sociocultural and constructivist views when learning mathematics (see Figure 1.2 on page 18). The publication of the National Research Council's report

Adding It Up: Helping Children Learn Mathematics (2001) summarized the most current research at the time on K–8 mathematics learning. Mathematical proficiency was multidimensional, and it continued the conversation about the need to understand mathematics and communicate students' thinking to depict proficiency. It was proposed that learning with understanding might be further enhanced by classroom interactions and Math Talks. Students who communicated their mathematical ideas and conjectures would learn to evaluate their own thinking and that of others, while developing more successful mathematical reasoning skills, leading to a deeper understanding of mathematics (Fosnot 2005; NCTM 2000). Once again, many mathematicians and mathematics educators found themselves at various places on this continuum, and debates provoked what is known today as the "math wars." This era of accountability continues to be a period of controversy. Paul Cobb's chapter in the *Second Handbook on Research in Mathematics Education* (2007) and NCTM's *How Students Learn: History, Mathematics, and Science in the Classroom* (2005) provide insights into the thoughts and theories that support the ideas of student-centered classrooms and encourage a culture of questioning, respect, and risk taking. These can be realized through the facilitation of mathematical discourse communities (National Research Council 2005, 13). In 2008, the National Mathematics Advisory Panel Report was published with the aim of ending the math wars. In 2010, with the release of the Common Core State Standards for Mathematics (CCSS-M), the emphasis was a balanced approach to support student understanding, reasoning, problem solving, and skill development (NGA and CCSSO 2010). The CCSS-M emphasized the Standards for Mathematical Practice (MP), which encourages all mathematicians to engage in habits of mind. These habits of mind include reasoning about mathematical strategies and critiquing the reasoning of others and are found when teachers and students engage in discourse. In 2015, as CCSS assessments were administered for the first time, we saw the first curriculum aligned with CCSS-M emerge. Within those curricula, discourse was included as a major emphasis to meet the rigor of thinking required for all students. At this point in time, NCTM's *Principles to Actions: Ensuring Mathematical Success for All* (2014) lead the way to carefully inform educators as they looked toward the obstacles that inhibited mathematics educators and the actions that define excellence.

Types of Discourse

The discourse of a classroom—the ways of representing, thinking, talking, agreeing and disagreeing—is central to what and how students learn about mathematics.

—NCTM 2007, 46

Mathematical discourse is the act of articulating mathematical ideas or procedures and can take place in several ways (see Figure 1.2 on page 22).

Levels of Discourse

Discourse takes on many different levels within one math lesson. These levels range from very simplistic to complex. See Figure 1.3. For example, student discourse that falls in levels 1 and 2 (answering and sharing) indicates a low cognitive demand and is considered lower quality discourse. Student discourse that is of high cognitive demand (justifying and generalizing) is considered higher quality discourse. Based on Sherin's (2002) research, Weaver and Dick (2006) created nine levels to document these types of discourse in their work, to help teachers reflect on and scaffold the levels of discourse that engage their students while reflecting on their practice.

Figure 1.3 Nine Levels of Discourse

Level	Definition	Explanation
1	Answering	A student gives a short answer to a direct question from the teacher or another student.
2	Making a Statement or Sharing	A student makes a simple statement or assertion or shares work with others, and the statement or sharing does not involve an explanation of how or why (e.g., reading what he or she wrote in a journal to the class).
3	Explaining	A student explains a mathematical idea or procedure by stating a description of what he or she did or how he or she solved a problem. The explanation does not provide any justification of the validity of the idea or procedure.
4	Questioning	A student asks a question to clarify his or her understanding of a mathematical idea or procedure.
5	Challenging	A student makes a statement or asks a question in a way that challenges the validity of a mathematical idea or procedure. The statement may include a counterexample. A challenge requires someone else to reevaluate his or her thinking.
6	Relating	A student makes a statement indicating that he or she has made a connection or sees a relationship to prior knowledge or experience.

Level	Definition	Explanation
7	Predicting or Conjecturing	A student makes a prediction or a conjecture based on understanding the mathematics behind the problem (e.g., recognizing a pattern in a sequence of numbers, making a prediction about what comes next in the sequence, or stating a hypothesis about an observed mathematical idea in the problem).
8	Justifying	A student provides justification for the validity of a mathematical idea or procedure by providing an explanation of the thinking that led to the idea or procedure. The justification may be in defense of the idea challenged by the teacher or another student.
9	Generalizing	A student makes a statement that is evidence of a shift from a specific example to the general case.

Adapted from Weaver and Dick 2006

Conclusion

Dominant cultural beliefs about teaching and learning mathematics from the nineteenth century continue to be obstacles for consistent implementation of effective teaching and learning in our classrooms. It is not necessary to choose between theories. A learning theory is not a teaching strategy, but theory informs teaching. The traditional teaching of mathematics asked students to work in silence as the optimal learning condition, but this has not been proven to be productive. Classroom discussions, based on students' ideas and solutions to problems, are absolutely "foundational to children's learning" (Wood and Turner-Vorbeck 2001, 186). Students need to talk through strategies to know whether they really understand them. When hearing strategies, they may seem to make sense, but explaining them to someone else is the best way to know whether they are deeply understood. When students are asked to engage in discourse, they see that they have responsibility for the direction of their learning.

In the chapters that follow, we will explore some specific teacher moves, such as Math Talks, developing growth mindsets, choosing demanding tasks, and asking genuine questions as part of pedagogical strategies that can engage all students in these higher levels of mathematical discourse.

Reflect and Discuss

1. Why is discourse important in today's math classroom?

2. What historical patterns continue to influence your use of mathematical discourse?

3. What types or levels of discourse do you most often use? Which levels might you want to work toward?

4. How can your knowledge of the types and levels of discourse assist you in promoting mathematical conversations in your classroom?

5. What are your personal goals for using discourse in your math classroom?

Chapter 2

Discourse and Mathematical Practice and Process Standards

For many of us, learning mathematics was a silent place where only the teacher talked as he or she modeled how to complete procedures to get to solutions. As students, we sat in rows and filled out worksheets of problems in isolation. The teacher could be heard saying, "Sh! Quiet! Do your own work!" Luckily, that is not what a mathematics classroom looks or sounds like today. Today, we believe that a math classroom is full of productive talk and a place where students work together to share ideas and engage in solving complex problems while puzzling and thinking deeply about them, often over many days. State and national mathematical practice and process standards support this type of classroom where all students are actively engaged in solving problems, using tools strategically, communicating their ideas, and reasoning about the ideas of others. Mathematical practice and process standards describe the habits of mind that mathematicians routinely practice. As math educators, we are always looking for ways to model, develop, and support mathematical habits of mind with all of our students. These standards provide a framework for deepening our teaching and our students' learning of mathematics in the classroom (see Figure 2.1 on page 27).

Mathematical practice and process standards are deeply rooted in the connections between teaching and learning. In 2001, the National Research Council's report, Adding It Up: Helping Children Learn Mathematics, defined mathematical learning to include the development of five interrelated strands that could produce mathematically proficient students. These five interrelated strands include, conceptual understanding, procedural fluency, strategic competence, adaptive reasoning, and productive disposition. The strands have been described as NCTM's (2000) process standards, reasoning habits

(NCTM 2009), and now mathematical practice (NGA and CCSSO 2010). In this chapter, we will refer to these practice and process standards as the mathematical habits of mind, which students need to develop as they grow in mathematical maturity and expertise. A purposeful focus on these habits of mind has the ability to support discourse and interactions between students and teachers in whole group lessons, small groups, student partners, and with the teacher. To remind, the purpose of discourse is not for students to state their solutions and get validation from the teacher but to engage all learners in thinking mathematically to maintain a high-cognitive demand throughout the lesson (Smith et. al 2009). Some of the most powerful discourse you will see is the mathematical conversations that occur between students nurtured by teachers.

Learning mathematics is an active learning process in which each student builds his or her own mathematical knowledge of personal experiences coupled with various forms of feedback from peers, teachers, other adults, as well as themselves. The National Research Council (2012) specifically states that learners should have experiences that enable them to:

- **engage** with challenging tasks that involve active meaning-making and support meaningful learning;

- **connect** new learning with prior knowledge and informal reasoning and, in the process, address preconceptions and misconceptions;

- **acquire** conceptual knowledge, as well as procedural knowledge, so that they can meaningfully organize their knowledge, acquire new knowledge, and transfer and apply knowledge to new situations;

- **construct** knowledge socially through discourse, activity, and interaction related to meaningful problems;

- **receive** descriptive and timely feedback, so that they can reflect on and revise their work, thinking, and understanding; and

- **develop** metacognitive awareness of themselves as learners, thinkers, and problem solvers and learn to monitor their own learning and performance.

Mathematical practice and process standards provide a framework for strengthening the teaching and learning of mathematics and strongly support how discourse can be a high-leverage practice to implement often and scaffold students' development of these habits of mind. Teachers have many tools available to them to understand what mathematical practices and processes seek to develop in all students, such as www.achievethecore.org, www.illustrativemathematics.org, www.insidemathematics.org, and www.nctm.org.

It is important, therefore, to deeply understand these standards. However, knowing what to look for, how to plan for, and how to scaffold student development of them is not always clear. One avenue has been to use a setup like the one below, which features the Common Core State Standards and was developed by Dr. Bill McCallum (2011). Here we see examples of what teachers should plan for in their lessons and how teachers may develop practices and processes for students. See Figure 2.1. This setup can support students in making connections between their actions as mathematicians and the math content they are exploring with more discourse opportunities.

Figure 2.1 Mathematical Practice

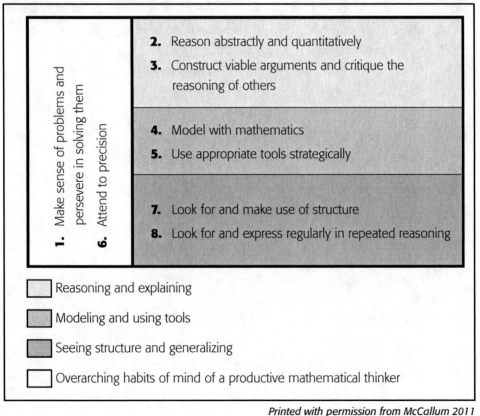

Printed with permission from McCallum 2011

So, what should teachers plan for in their lessons? How will they know that students are engaging in mathematical practice? What questions can teachers ask to scaffold students' learning using habits of mind? In the summer of 2011 at the National Council of Supervisors of Mathematics (NCSM) Summer Leadership Academy in Atlanta, GA, a draft document was created to support the Standards of Mathematical Practice. It contained a chart of teacher actions and student dispositions for each of the standards. Many school districts, teachers, and coaches have used this draft to create look-for and planning tools to support teachers, parents, and students when making the connections between content standards and practice standards. There are many resources and outlines for process standards as well. The following adapted list in Figure 2.2 can assist teachers in these endeavors of guiding future productive mathematicians.

Figure 2.2 Overarching Habits of Mind for a Productive Mathematician

Problem Solving
MP1 Make sense of problems and persevere in solving them.   **MP6 Attend to precision.**
Teachers plan for: • rich, cognitively demanding problem-based tasks that encourage students to persevere to reach a solution over time;   • tasks that allow for multiple entry points and solution paths, as well as, various representations: visual models, real life situations, diagrams, manipulatives, and symbols to make connections among the various ways students solve problems;   • tasks that require students to justify, argue, clarify, and defend solutions; and   • tasks that are relatable to the student populations, so they can access and make use of their relevant prior knowledge and experiences while understanding the tasks and working through them.

Teachers look for students who:

- take time to deeply understand the problem and look for entry points;
- make conjectures and plan a route to prove or disprove them;
- analyze the mathematical quantities/representations and think about their relationships;
- self-monitor and evaluate their progress and make changes as needed, including asking themselves if what they are doing makes sense;
- communicate precisely, orally, in writing and/or with numbers;
- calculate efficiently and provide carefully formulated explanations with accurate labels; and
- express solutions within context of the problem or the strategy used to get to the solution.

Sample teacher questions to elicit the desired behavior:

- What do you notice about…?
- What information do you think is important? How do you know?
- Tell me what you have already tried. What could you change?
- What are you most confident about so far?
- How would you describe this problem in your own words?
- What might you do to get started?
- Is there anything you know for sure about this problem that would help?
- Does your answer seem reasonable? How do you know?
- Is there a tool that would help you find a precise solution?
- What can you do to double-check your solution?
- How could you test your solution?

MP2 Reason abstractly and quantitatively.
MP3 Construct viable arguments and critique the reasoning of others.

Teachers plan for:

- tasks that ask questions and that require students to attend to the meaning of the quantities and their relationships, not just how to compute them;
- tasks that use relevant, meaningful content for the current student population;
- tasks that expect students to show their thinking using symbols and to explain the pathway to their solutions;
- tasks that embed discussion and communication of reasoning and ask for justifications of strategies;
- tasks that require students to provide evidence to explain their thinking beyond merely using computational skills to find a solution and to share their ideas with others; and
- discussion circles that give all students the opportunity to ask clarifying and challenging questions of each other to provide feedback and scaffold learning.

Teachers look for students who:

- make sense of and explain quantities and relationships in problems;
- make connections to prior knowledge, both mathematical and real-life;
- translate symbolic equations into context to visualize and reason logically about them;
- question others about their thinking and clarify their own understanding;
- form logical arguments using conjecture and different thinking;
- build new ideas from other students' ideas;
- ask clarifying and challenging questions of other students' strategies, solutions, and ideas; and
- use multiple representations, tools, and/or models to give evidence of their thinking.

Reasoning and Explaining *(cont.)*

Sample teacher questions to prompt discourse:

- What do the numbers in this problem represent?
- What is the relationship of _____ and _____?
- How did you decide that you needed to use _____?
- Could we have used another strategy or operation to solve this problem? Why or why not?
- Will it still work if…?
- How did you figure it out?
- Does anyone else have another way of thinking about this problem?
- How are these strategies alike or different?
- Can you convince us?
- Can you prove that?

Modeling and Using Tools

MP4 Model with mathematics.
MP5 Use appropriate tools strategically.

Teachers plan for:

- tasks that invite students to connect to a real-world situation (context) that explains the mathematics;
- tasks that ask students to apply the mathematics they know to solve novel problems that arise in everyday life, society, and the workplace;
- tasks that ask students to apply what they know and make assumptions and approximations to simplify complicated situations;
- activities that lend themselves to the use of multiple learning tools;
- tasks that ask students to determine and use appropriate tools for a particular problem;
- activities that use materials, models, tools, and/or technology-based resources to make conjectures or be used to find solutions; and
- tasks that ask questions about students' choice of tools and ask students to solve their thinking with the chosen tool.

Teachers look for students who:

- apply prior mathematical knowledge to solve real-world problems;
- mathematize situations using numbers, symbols, equations, tables, graphs, or formulas;
- identify important information needed to solve a contextual problem and relate that problem to everyday situations;
- make sense of symbols or quantities in an equation or function (as they relate to context/real-life situations);
- make sound decisions about their choices and use of specific tools; and
- use technology to explore mathematical situations, to search for digital content to pose or solve problems, and to explore and deepen understanding of mathematical content.

Sample teacher questions to prompt discourse:

- Can you make a sketch to show your thinking?
- What are some ways you could represent the quantities?
- What equation could we use to represent this situation?
- How might you use _____ (tool) to show this situation?
- How does this relate to the real world?
- What is this situation about? Tell me in your own words.
- What approach are you considering trying first?
- Is there a math tool to help you see the problem?
- Why was it helpful to use _____ (tool)?

MP7 Look for and make use of structure.
MP8 Look for and express regularity in repeated reasoning.

Teachers plan for:

- tasks that ask students to look for patterns or structure, recognizing that quantities can be represented in different ways;
- activities that expect students to recognize and identify structures from previous experiences and apply their understanding in a new situation;
- tasks that connect and allow students to see complicated quantities as single objects or compositions of several objects and then use this understanding to make sense of problems;
- activities that give students opportunities to reveal patterns or repetition in thinking to challenge them to make a generalization or rule;
- tasks that require students to search for patterns and relationships to develop a mathematical rule; and
- tasks that expect students to discover the underlying structure of a problem and come to a generalization.

Teachers look for students who:

- make connections between skills and strategies previously learned to solve novel problems and tasks;
- search for, identify, and use mathematical patterns and structures to solve problems and make generalizations;
- see complex problems and simplify them by breaking them down into manageable chunks to successfully solve them;
- generate rules from repeated reasoning and previous mathematical practice; and
- evaluate the reasonableness of their strategies throughout the problem-solving process.

Sample teacher questions to prompt discourse:

- What do you notice?
- Do you see any patterns here?
- What might come next? Why?
- What do you predict will happen? Why?
- How is this problem like the one we just solved? How is it different?
- Does it always work? Why or why not?
- What would happen if…?
- How does this relate to…?
- Explain how this strategy would work in a new situation.

Adapted from NCSM 2011

Through the use of discourse in any mathematics classroom, teachers can plan their lessons and carefully choose tasks centered on engaging students in discussing meaningful mathematics to promote reasoning and problem solving. Teachers who strongly commit to this course of action learn to plan lessons that prompt student interactions and discourse with the goal of helping every student make sense of the mathematical concepts and procedures at hand. Focused lesson planning, careful choice of rich tasks, and lesson outcomes that keep the standards for practice and process in mind can be positive steps toward developing student-to-student discourse in any grade-level classroom.

Developing Growth Mindsets Using Discourse

In 2006, Dr. Carol Dweck, Professor of Psychology, Stanford University, published a book called *Mindset: The New Psychology of Success*. In her book, she shared over 30 years of research about how people succeed and her simplistic yet rich theory about two mindsets. Here is where the notion of a fixed mindset and a growth mindset began (Dweck 2006).

Fixed Mindset	Growth Mindset
The belief that we're born with a fixed amount of intelligence and ability. People operating in the fixed mindset are prone to avoiding challenges and failures, thereby robbing themselves of a life rich in experience and learning. —Dweck 2006, 15–16	The belief that with practice, perseverance, and effort, people have limitless potential to learn and grow. People operating in the growth mindset tackle challenges with aplomb, unconcerned with making mistakes or being embarrassed, focusing instead on the process of growth. —Dweck 2006, 16

These opposing mindsets—fixed and growth—exist in all of us, and our choices in our lives through the growth or fixed mindset make a big difference in building our efficacy as mathematicians. Dr. Dweck shares that we are all born with a growth mindset. Babies do not give up trying to talk when they make no sense. When they learn to walk, they fall down and get right back up!

So, when does this love of learning by making mistakes end? That was the question that Dweck asked and answered in her research. She concluded that, "As soon as children become able to evaluate themselves, some of them become afraid of challenges, they become afraid of not being smart" (Dweck 2006, 16). She also shared in her recent article that we all must legitimize the fixed mindset. As perpetual learners, we need to acknowledge that we're all a mixture of fixed and growth mindsets and we probably always will be. If we want to move closer to a growth mindset in our thoughts, practices, and learning, we need to stay in touch with our fixed-mindset thoughts and deeds (Dweck 2015). If one carefully watches for our fixed-mindset triggers, we can begin the true journey to a growth mindset. Teaching students to watch for fixed-mindset reactions when facing challenges, accepting those thoughts and feelings and choosing to work through them are signs of a true growth mindset. Dweck shares examples, which I have expanded to document the growth mindset at work throughout history in some very famous and successful people. Here are a few growth mindset stories to ponder.

Marie Curie (Pasachoff 2017) was born in war-torn Poland where women were not allowed to pursue higher education. She found ways to create opportunities to learn about math, chemistry, and physics, the subjects she loved at great personal risk. Later, Curie became the first woman to win a Nobel Prize. Albert Einstein, too, wasn't able to speak until he was almost four years old, and his teachers said he would "never amount to much." The Beatles were rejected by Decca Recording studios who said, "We don't like their sound—they have no future in show business." Walt Disney was fired from a newspaper for "lacking imagination" and "having no original ideas." Steve Jobs, at 30 years old, was left devastated and depressed after being unceremoniously removed from the company he started. Oprah Winfrey was demoted from her job as a news anchor because she "wasn't fit for television." And, in her book, Dweck shares about many sports heroes who succeeded from their growth mindset. One of them was Michael Jordan, who was cut from his high school basketball team. Afterward, he went home and locked himself in his room and cried. Yet, one of Michael Jordan's most often used quotes supports his growth mindset:

> **If you're trying to achieve, there will be roadblocks. I've had them; everybody has had them. But obstacles don't have to stop you. If you run into a wall, don't turn around and give up. Figure out how to climb it, go through it, or work around it.**
>
> **—Michael Jordan**

The growth mindset of all these historical figures was their road to success! So, how do we make this a reality in the classroom?

Dr. Jo Boaler, Professor of Mathematics at Stanford University, has worked for many years with her colleague, Carol Dweck, and their collaboration and research projects were born because they both agreed that mathematics was the subject most in need of a mindset makeover. Dr. Boaler (2016) takes this idea and applies it to learning mathematics in her numerous publications and is committed to teaching children through the education of teachers, parents, and the community at large to develop mathematical growth mindsets. Building a mathematical mindset community of learners is probably the single most important way to create a discourse-rich classroom. Teachers can begin by building strong relationships with individual students in their classrooms.

These six beliefs have to be in place when building student-to-teacher relationships, which will allow discourse to be prevalent at any grade level:

- Students know that their teacher has high expectations for every individual.

- Students know that their teacher believes in their ability to learn.

- Students respect and admire their teacher as a person.

- Students seek and desire their teacher's feedback.

- Students know that growth is more important than correct answers and grades.

- Students know their classroom is a safe environment to take risks.

As this environment is established, the teacher, alongside students, must create the norms of a mindset community of learners. Using Jo Boaler's (2017) seven positive norms to encourage risk-taking in math class combined with the Four Freedoms (see page 38) of a successful math classroom, which have been posted in my classroom for years, any teacher can build community as the first step in creating a discourse-rich classroom.

Positive Norms to Encourage in Math Class

High expectations for all students in math class are required and expected by the teacher. We must encourage students to believe in themselves. There is no such thing as a "math" person. We were not born with "math genes." Everyone can reach the highest level of mathematical learning with hard work and perseverance.

Mistakes Are Valuable

Mistakes give you opportunities to grow your brain. It is good to struggle with mathematical ideas and make mistakes. Make a new mistake daily to continue being a lifelong learner! The key word is new! Risk-taking is expected.

Questions Are Really Important

Always ask questions. Always search for the answer to your questions. Ask yourself: Why does this make sense? Why does that work? How do I know? What do I already know that I can use? What do I wonder right now?

Math Is about Creativity and Making Sense

Math is a very creative subject that is, at its core, about searching for and visualizing patterns and relationships. We are all looking for and creating solution paths that others can see, discuss, clarify, and critique. Students use ownership moves and terms, such as *my method* or *my thinking*, to be open to other ways of thinking.

Math Is about Connections and Communicating

Math is a connected subject and a form of communication. Being able to talk about ideas, strategies, models, and thinking is expected. Representing your ideas in different forms deepens your thinking. Link your words, pictures, graphs, and equations together to effectively communicate ideas. Use your prior knowledge to connect to new problems and learning.

Depth Is Much More Important Than Speed

Top mathematicians think slowly and deeply. Cathy Twomey Fosnot wonders, *Where did the notion that mathematicians are fast come from? We are slow and methodical and want to puzzle about ideas and engage in conversation with other mathematicians about our wonderings.*

Math Class Is about Learning Not Performing

Math is a growth subject. Be open to learning something new every day. We learn over time by constantly exerting effort and perseverance. The math classroom should be filled with wonder and curiosity. It is awesome to say, "I never thought of it like that before."

The Four Freedoms in Room 16

Learning often takes its own pace. We, as teachers, cannot always dictate the moment a student "gets it." But, students can surprise themselves when they are introduced to the four freedoms of learning (Ronda 2012). These freedoms, which I have had posted in my classroom for years, can have a lasting impact on student cognitive development and critical thinking. They can be opportunities to teach students mental math, number systems, and symbols without students even realizing they are internalizing them.

The Freedom to Make Mistakes

Help students approach the acquisition of knowledge with confidence and perseverance. We all learn through our mistakes. Listen to and observe your students and encourage them to explain or demonstrate why they think what they do. Support them whenever they genuinely participate in the learning process. If your class is afraid to make mistakes, they will never reach their full potential.

The Freedom to Ask Questions

Remember that questions students ask should not only help assess where they are but should also assist when evaluating our own ability to foster learning and growth. A student showing an honest effort must be encouraged to seek help. We will all learn from each other's questions. The responses we give depend upon the student and the question, but we should never make the student feel the question should never have been asked.

The Freedom to Think for Yourself

Encourage your class to reach their own solutions. Do not stifle student thinking by providing solutions, algorithms, or your ideas before allowing each student time and opportunity to feel the rewarding satisfaction of achieving a solution on their own. Once we know that we can achieve, we may also appreciate seeing how others reached the same goal. Give students the freedom to think.

The Freedom to Choose Your Own Methods

Allowing students to select their own paths guided by knowledge and thinking will help them realize the importance of **thinking** about the mathematics rather than trying to **remember** it.

Many teachers and students are working together to change their words and thoughts as they build growth mindsets. Some create posters or bulletin boards with phrases or thoughts to change and what to change them to as shown in Figure 2.3 on page 40:

Figure 2.3 Phrases and Thoughts to Change Your Mindset

Change Your Words	to	Change Your Mindset
I'm not good at this.	➡	What am I missing? I am just not there YET!
This is too hard.	➡	This may take some time and effort.
I made a mistake.	➡	I can learn from my mistake.
I will never be as smart as her.	➡	I am going to figure out what she did and try it!
I can't do math.	➡	I am going to train my brain in math.
I give up.	➡	I will use a strategy I have learned.
It's good enough.	➡	Is this really my best work?
I can't make this any better.	➡	I can always improve and will try another way.
I'm awesome at this!	➡	I am on the right track and can do more.

Focusing on student mindset and thinking about your own mindset when it comes to teaching mathematics is a powerful tool for supporting discourse in the classroom. If teachers truly believe that every student has the power to grow and improve his or her talents, skills, and abilities with hard work and effort, and truly desires to cultivate that belief in all students, the seeds of a growth mindset will be planted. Then, students will be empowered as thinkers and problem solvers for years to come.

Productive Student and Teacher Roles

Mathematical discourse among students is critical for the depth of learning expected of mathematicians in the twenty-first century. When teachers carefully plan their whole group and small group lessons to intentionally facilitate discourse built on ideas, strategies, and thinking, they guide the learning of the class in a productive, disciplinary direction. Students need to expect that they will all be active members of the discourse community as they explain their reasoning and consider the mathematical explanations and strategies of their classmates. The teacher and students have an interactive role in this facilitation, but the goal for the teacher is to facilitate the student-to-student discourse by orchestrating genuine questions, arguments, and thoughts to empower all students as mathematicians. The following roles in Figure 2.4 can guide both teachers and students as they plan for and engage in meaningful discourse in the mathematics classroom.

Figure 2.4 Facilitating Mathematical Discourse in the Classroom

---

Teacher's Role

- Set lesson goals that identify the big math idea and what students are meant to understand about mathematics.

- Choose rich, cognitively demanding tasks that demand engagement with concepts and stimulate students to make connections.

- Create parallel tasks that give multiple access points to the problem to ensure learning for all students.

- Select and sequence student approaches and solution strategies for whole-class analysis and discussion.

- Engage students in purposefully sharing mathematical ideas, reasoning, and approaches using varied representations.

- Facilitate discourse among students by positioning them as authors of ideas who explain and defend their approaches.

- Ensure progress toward mathematical goals by making explicit connections to student approaches and reasoning.

Figure 2.4 Facilitating Mathematical Discourse in the Classroom *(cont.)*

Student's Role

- Set personal goals toward the lesson outcome and know it is your responsibility to share ideas, partial thinking, strategies, and/or solutions.

- Know that it is everyone's responsibility to listen to others' ideas and ask clarifying questions.

- Choose tasks that will challenge your thinking and allow you to puzzle them.

- Be prepared to present and explain ideas, reasoning, strategies, and models that represent your ideas to one another, in pairs, small-group, and whole-group discourse.

- Work to actively listen to and critique the thinking of your peers; using ideas, arguments, or questions to clarify the thinking and ideas of others.

- Strive to understand approaches other than your own by asking clarifying questions and trying out differing strategies other than your own.

- Describe and identify how various approaches to solving a task are the same and how they are different.

When orchestrating productive discourse in the K–12 classroom, Smith and Stein (2011) share the decisions that teachers need to make when using the following five practices for effectively using student responses in classroom discussions:

1. **Anticipating** student responses before the math lesson begins
2. **Monitoring** student work during their lesson and student engagement with planned tasks
3. Carefully **selecting** student strategies/mathematical approaches to present
4. **Sequencing** student work in a specific order to scaffold discussion and learning
5. **Connecting** different student ideas, strategies, and arguments to key mathematical ideas/concepts

Productive discourse is not an accident, and teachers cannot work completely on the fly, hoping student discussions will suddenly become important mathematical ideas and learning. Accomplishments take classroom community. It is a driving force in how students learn as a group. And teachers who plan, keeping their roles in mind while using effective practices, will find themselves in control of productive classroom discussions.

Culturally and Linguistically Responsive Teaching and Learning: Building Receptive Language Functions for All Students through Discourse

Often, teachers have the misconception that mathematics is not associated with any language or culture because it uses symbols and numbers. But, if we really think deeply about it, language plays a key role in doing, learning, and making sense of mathematics. As teachers, we use language to explain new mathematical concepts and to strategically carry out procedures and formulas while using specialized vocabulary. Culturally and Linguistically Responsive (CLR) teaching and learning means that a teacher chooses to transform his or her practice to ensure that all students are included. Dr. Sharroky Hollie's *Strategies for Culturally and Linguistically Responsive Teaching and Learning: Building Receptive Language for All Students* (2015) challenges teachers, administrators, and students to incorporate students' home language into the classroom, and his approach can be used in mathematics as well. When teachers develop responsiveness to their students' culture and learning, they validate and affirm "the (home) culture and language for the purpose of building and bridging the student to success in the culture of academia and in mainstream society" (Hollie 2015, 13). When solving mathematical problems, we use a receptive language function like this to make sense of the problem and reason about it. The use of discourse has been seen to deepen student understanding through conversations, reasoning about other student's ideas, and learning to think in different ways. When students share ideas, argue, defend their strategies, and genuinely question each other using familiar and academically minded language, they begin to use and understand the language of mathematics. Some of the obstacles of mathematical language acquisition are 1) unknown, misunderstood, or confusing words, 2) difficulty comprehending mathematical problems and relating them to their own background experiences or prior knowledge, and 3) the linguistic demands within a mathematical lesson.

Bresser, Melanese, and Sphar (2009) share a partial list of tricky math words that may interfere with student mathematical comprehension, yet they must be able to comprehend, understand, and incorporate them into their language and learning. See Figure 2.5 on page 44. It is our job to embed excellent teaching strategies to help students make sense of these tricky academic math words. What do you notice about this partial list of terms?

Figure 2.5 Tricky Words in Mathematics

acute	inscribe	rational
altitude	intersection	ray
base	irrational	reflection
brackets	irregular	regular
change	key	relative
closed	left	right
combination	mass	root
composite	mean	round
coordinate	median	ruler
count	multiple	scale
degree	negative	segment
difference	net	set
digit	obtuse	side
edge	odd	similar
even	open	slide
expression	operation	solution
face	origin	space
factor	period	sum
fair	plane	table
figure	plot	term
foot	point	times
formula	power	translation
function	prime	union
identity	product	unit
if	proper	value
improper	property	volume
inequality	range	yard

The words above have multiple meanings and can be easily misunderstood. Some of the words have the same meaning. For these reasons, it is important to teach academic vocabulary in context as often as possible. What follows is a lesson entrenched in discourse opportunities that can help students connect prior knowledge to vocabulary while engaging them in a paperfolding lesson using a simple circle. This lesson was written by Linda Foreman and Albert B. Bennett Jr. It is from a lesson in *Visual Mathematics* (1996) and adapted with permission from the Math Learning Center.

Paperfolding

Predicting and examining the results of paperfolding provide a context for students to:

- communicate about mathematical relationships;

- strengthen spatial sense;

- develop intuitions about symmetry;

- apply understanding of angle measurement and angle relationships; and

- deepen understanding of geometric shapes and the relationships among shapes.

1. Cut out the circle on Connector Master A in Figure 2.6 on page 46 and as a reproducible on the Digital Resources (paperfolding.pdf). Fold the circle so the three points on the circumference coincide with the center of the circle. (The folded figure is technically a triangular region, and its boundary is an equilateral triangle. However, it is common practice to refer to the shape as an equilateral triangle. Similarly, the remaining actions of this activity will produce regions, but we will discuss only the perimeter of the regions.)

2. Ask students to work in groups to discuss and list conjectures about attributes of the folded figure they formed and develop informal methods to verify their conjectures. Ask for volunteers to share methods and observations. List all the geometric vocabulary mentioned on a chart or whiteboard.

3. Ask students to locate and mark (by creasing lightly) the midpoint of one side of the triangle. Then, have them fold the opposite vertex to that midpoint to form a trapezoid. Again, discuss and list conjectures about the attributes, and list the geometric vocabulary on a chart or whiteboard.

4. Fold one of the end triangles over the middle one to form a rhombus (parallelogram). Discuss the results, and continue to record the geometric vocabulary.

5. Fold the other end triangle over the middle one to form a smaller equilateral triangle, and discuss the results. How many similar (smaller) triangles make up the original larger triangle? What terminology do we use when shapes are the same shape but different sizes? Record the geometric vocabulary students use.

6. Open to the original triangle. Allow the three smaller triangles to come to a point to form a 3-dimensional figure known as a tetrahedron. Discuss and record vocabulary.

7. Open to the original triangle and fold the three vertices to the center point to form a regular hexagon. Discuss and record vocabulary.

8. Lift the vertices of the three small triangles that folded over to form the hexagon and carefully push them toward the center to form the 3-dimensional figure called a truncated tetrahedron. Discuss and record vocabulary.

9. Unfold the circle, and identify the polygons formed by the creases. Use colored pencils to find as many polygons as possible.

10. Find a geometric vocabulary term, and fold your circle into a shape that allows you to demonstrate the meaning of the term, and post it next to that word.

The academic vocabulary introduced and reinforced in this activity provides a basis for much geometric exploration. Some possible journal entries or sentence frames at the conclusion of the lesson could be:

- "Here is everything I know about the _____(name of shape)."

- "The _____(shape name) and the _____(shape name) are similar/ different because _____."

"If math was a shape it would be…"

List three important geometry words, and use pictures, words, and/or numbers to define it.

Adapted from a lesson in Visual Mathematics with permission from The Math Learning Center Salem, Oregon

www.mathlearningcenter.org

Figure 2.6 Connector Master A

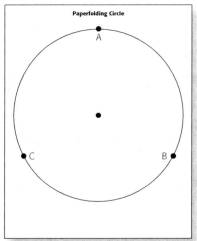

This lesson models how language can be embedded in a lesson rather than taught out of context at the beginning of a lesson. As teachers, we can support all of our learners as they develop precise mathematical vocabulary. Engaging all learners in discourse is one of the key ways to support English language learners and native speakers with well-designed support, so they can develop proficiency in English and deepen their mathematical understanding at the same time. It is important to talk about and write math to learn math. Bresser, Melanese, and Sphar (2009) share many strategies that support learning English in the classroom. Figure 2.7 lists strategies adapted from their ideas that specifically support discourse and mathematical learning in the K–12 classroom, which teachers should consider incorporating into their lessons.

Figure 2.7 Strategies That Support Learning Math While Learning English

Making Math Content Comprehensible

- Create vocabulary banks with photos/pictures. (See word resource app at www. mathlearningcenter.org/resources/apps/math-vocabulary-cards.)
- Use native language as a resource, especially in small group instruction when scaffolding learning.
- Make manipulative materials and tools available at all times.
- Activate prior knowledge and student background experiences.
- Provide meaningful visuals, including real items.
- Pose problems in familiar contexts.
- Demonstrate and model (both teachers and students).
- Use dramatization and gestures.
- Modify teacher talk.
- Revoice mathematical ideas and terms.
- Connect symbols with words/actions.

Providing Opportunities for Talk

- Ask genuine questions that elicit explanations from students, not just solutions.
- Consider both language and math competency when grouping students.
- Facilitate whole-group discussions by asking high-leverage, genuine questions.
- Allow for small-group or partner discussion before a whole-group discussion.
- Utilize partner talk as often as possible.
- Change table partners weekly to vary the level of talk and mathematics.

Supporting Talking
• Use sentence frames or sentence starters for journaling prompts and in lessons where students may need support with the mathematical language.
• Practice wait time. (Be patient and give all students time to think deeply.)
• Design parallel tasks and ask questions that target different proficiency levels.
• Provide a safe environment where students can take risks.
• Establish mathematical community agreements or norms.
• Elicit nonverbal responses (e.g., thumbs up, response cards, signals).

Adapted from Bresser, Melanese, and Sphar 2009

Many of these strategies support the development of both productive and receptive language function for students. Getting students to produce language appears to be easier than developing receptive language in individual students. Building receptive language functions when teaching mathematics is complex. Receptive language is the understanding of language input. This goes beyond just vocabulary skills and includes the ability to interpret a question as stated. For example, students need to understand conditional statements, such as "If the early bird rises, then it gets the worm," which means that because the bird prepared early, it had a better chance of being successful; as well as, accurately interpreting complex grammatical forms, such as "The girl was licked by the dog," which means that a dog did an action—the dog licked a girl.

When solving mathematical problems, students need to comprehend the meaning of a problem as presented in multiple representations, such as spoken language, written texts, diagrams, drawings, tables, graphs, and mathematical expressions or equations. This can be tricky enough for native language speakers let alone for students learning a second language. They must successfully comprehend others' talk about math problems, solutions, approaches, strategies, and reasoning. Then, they must coordinate these various formats and multiple representations of their peers' ideas to make sense of the mathematics. This can be a very complex process and has to be intentionally integrated into all math lessons.

One of the most successful protocols for solving problems to engage all students in mathematical discourse was introduced to me by Dr. Harold Asturias of the Lawrence Hall of Science in Berkeley, CA in April 2012. He shared an adaptation of Grace Kelemanik's *Three Reads* that specifically develops receptive language function while addressing the eight Standards for Mathematical Practice via the use of mathematical discourse. Through the use of this routine, second language learners are given a structure that supports their comprehension, thinking, reasoning, and justification when solving any complex word problem.

We know that students at all levels struggle with reading and restating story problems in their own words, which limits comprehension of what the problem is about. Most often, students underline the numbers and look for one word that might help them know which operation to use in solving the problem. They struggle with identifying the correct operation because they did not step back and think deeply about the context of the story problem. This method of listening for or highlighting one keyword is often explicitly taught as a problem-solving method. Yet, this keyword approach encourages students to ignore the meaning and structure of the problem and look for the quick and easy way out. It often leads students to think that solving problems quickly is more important than making sense of the problem and attending to precision. Students do not try to visualize the structure or see the patterns in the problem, therefore, when a problem is complex, they struggle with breaking it down into simpler, logical steps or sequences. Frequently, they do not think about what they know or apply prior knowledge of strategies correctly. They often feel that if they make a conjecture or guess about how to start a problem that they must not be "good" at math. Finishing first is what they think matters, and they rarely attempt to justify their thinking or solutions. We know that accurately retelling a story in your own words and bringing prior knowledge to stories is a sign of excellent comprehension in reading. So, why not bring it into mathematics class as well?

The good news is that, at any grade level, teachers can teach their students to use the Three Reads problem-solving routine with any complex math problem. Within these three readings, students are listening and reading with a particular focus. They engage in student-to-student discourse within the three simple readings. The process begins with students taking time to comprehend the context of the story problem (text), which is presented as a problem stem. A problem stem is simply the situation for a math word problem with the question removed. Secondly, students are asked to identify and think about the relationships of the mathematical structure of the situation. Thirdly, they brainstorm a list of all the possible mathematical questions they have about this contextual situation and sketch a diagram using their understanding of the context and the mathematics to devise a plan to answer one of their posed questions. These three steps begin the problem-solving process for a whole group lesson. Then, students are given time to work on the problem, which usually happens over several days, depending on the complexity of the problem to prepare for a future class discussion. The daily activities are aligned to support mathematical practice and process standards found in state and national standards. CCSS's Standards for Mathematical Practice (MP1–MP8) and the Texas Essential Knowledge and Skills for Mathematics (TEKS-M) Mathematical Process Standards (A–G) are shown as examples.

A Sample Three Reads Lesson— Hannah's Problem

Beginning the Lesson:

Today, we are going to work together on making sense of a mathematical problem. You may have solved math word, or story, problems before, but today we are going to be readers in math class. Just like when we read stories in books, we have to take time to comprehend the story and think what we know about this type of situation and how it will help us understand what is happening in the story. We have to do the same thing with a math story problem, as well. We will read the story problem three times because good readers reread stories to make sense of them. Each time I am going to ask you to discuss and respond to a specific question. The first time you have to listen very carefully because I am going to read it aloud to you. You won't have it in front of you. So, be a good listener and think about what is happening in the story. How could you retell the story in your own words?

First Read Focus: Comprehending the Context of the Text

Discuss: Ask students what the situation is about to make sense of the problem (MP1 and TEKS-M 1.B).

Listen carefully to our story. Good readers are good listeners and should always listen for the context of the story to make sense of it in their own minds.

The teacher should use this read to make sure students focus on the context of the story only, not the mathematics, numbers, or answers. Teachers check for student understanding of the words and vocabulary that describe the situation.

The teacher can read the problem stem (a word problem without the question) aloud. **Note:** Avoid posting or showing students the written problem to give them opportunities to actively listen.

Sample Problem: Hannah's age this year is a multiple of five. Next year, Hannah's age will be a multiple of four.

Discuss: Encourage student-to-student discourse as they make sense of the problem (MP1 and TEKS-M 1.F) and reason abstractly about the problem (MP2 and TEKS-M 1.B).

How would you describe the problem in your own words? What is the context of the story? Can you retell the story using your own words?

Students interpret the abstract and concrete objects in the text while making sense of the quantitative relationships presented to them. By actively listening, students can understand their peers' approaches for interpreting these relationships, which may be different from their own.

Second Read Focus: Comprehending the Mathematics

Discuss: Encourage student-to-student discourse, using think-pair-share or whole class discussion (MP6, MP7, and TEKS-M 1.D).

What do you notice about the quantities? (MP6 and TEKS-M 1.G) Or, are there any patterns or relationships (MP7 and TEKS-M 1.F)?

We have talked about the situation in the story. Good readers listen for important details. They think about their prior learning and background experiences, also known as prior knowledge, and use it to make sense of the story.

Display the word problem for all to see.

I will read the story again, and I want you to follow along with me as I read, so we can think about the mathematics in the problem.

Post the problem so students can follow along as the teacher reads aloud.

What quantities, the numbers and their units, are important, and what relationships are there between those quantities? Mathematicians look for explicit quantities (e.g. multiple of 4) and implicit quantities (e.g., next year) when they are given a problem to solve. Always look for both! Sometimes they are not numbers!

What are the quantities in this situation? How are those quantities related (MP2 and TEKS-M 1.F)?

Questions will elicit a need for precision and making sense of the structure (MP and TEKS-M 1.G). **Note:** Asking how the quantities are related is very important in this discourse. This is the high level questioning that mathematicians ask to make sense of the problem.

Third Read Focus: Listing All Possible Mathematical Questions

Discuss: Have students engage in student-to-student discourse by using the think-pair-share strategy to form questions for the problem (MP1 and TEKS-M 1.B).

Now, you will read the story problem silently to yourself one more time. As you read, think as a mathematician would about all the possible questions we could ask about this situation. Do not look at what the people in the situation are wearing or doing, but pose questions about the quantities and their relationships.

What are all the possible mathematical questions we could ask of this situation (MP1, TEKS-M 1. A, and TEKS-M 1.F)?

Teachers can make a list on the whiteboard or a chart of students' brainstormed questions. Do not remove any, but record them all, even if students cannot answer the questions with the given information. Students will figure out which questions they can answer and which they can't answer as they solve the problem. It is also important to never tell students what the book's questions were. We want this process to give ownership to each student's thinking and question generating.

For example, Mrs. Laursen's fifth grade class posed these questions:

- How old is Hannah now?
- How old will she be on her next birthday?
- Can Hannah be a fifth grader?
- Can Hannah have a driver's license?
- Can Hannah be my grandma?
- What does Hannah look like?

Working on the Problem

Here is a copy of the problem for each of you to put in your math journal.

Provide students with the problem stems (no questions). Place mailing labels or cut strips to be glued into student journals.

Sample Problem:

Hannah's age this year is a multiple of five. Next year, Hannah's age will be a multiple of four.

Before you start calculating and working toward a solution, I want you to draw a diagram to help you work toward solving the problem. A diagram is a tool that mathematicians use to represent a problem situation, so they can better understand a math problem. Make sure your diagram shows the context of the story, the quantities in the story, and how they are related (MP4 and TEKS-M 1.C). As you sketch your diagram of what you know, I will look for students to share their diagrams. Then, you will have several days to work and prepare your ideas, arguments, and questions for our math forum this Friday. So, use the next few minutes to sketch your diagram.

Discuss/Teacher-to-Student Discourse: Compare and contrast different students' ways of thinking (MP2, MP3, MP5, and MP6; TEKS-M 1.A, TEKS-M 1.B, TEKS-M 1.D, TEKS-M 1.F, and TEKS-M 1.G).

This is mathematician _____'s diagram. Who can explain _____'s ideas or thinking (MP3, TEKS-M 1.D, and TEKS-M 1.G)?

How is mathematician _____'s way of thinking similar or different than yours (MP2 and TEKS-M 1.F)?

The teacher carefully selects two diagrams to share.

How are these diagrams the same? How are they different (MP2, TEKS-M 1.D, and TEKS-M 1.G)?

Throughout this process, the teacher carefully selects and sequences the diagrams from simple situational context diagrams to more complex diagrams using the mathematical quantities that show how they relate to the meaning of the problem.

Talk to your partner about his or her diagram. What do you see in it that helps you make sense of the problem at hand? Is there any information that you might want to add to your diagram to clarify your thinking at this point (MP3 and TEKS-M 1.B)?

Now, choose a question from this list or maybe even a new one that you have formed from our exploration of diagrams, which you are interested in answering.

1. Persevere and answer another question.
2. Justify your solution using words, numbers, and/or pictures.

3. Can you prove your solution is correct? Can you get to your first solution using another strategy?

4. Do you and your partner have different strategies but got to the same answer?

5. Are there other questions on our list that you can answer? Do you have any questions that you want to add to our list?

Other possible questions or coaching prompts to engage students in discourse during student diagrams:

1. Which diagram best helps you see and make sense of the problem? Why?

2. Can you make a connection between these two diagrams?

3. Where do you see the relationships between the quantities in this information?

4. Do any of you have any clarifying questions you would like to ask of mathematician _____?

5. How does this diagram show_____? (Name a specific quantity and/or relationship.) (MP2, MP3, and TEKS-M 1.F)

6. Where is_____? (Name a specific quantity and/or relationship in the diagram.) (MP2, MP4, C, and TEKS-M 1.F)

7. Do you see any new relationships in this diagram_____? (One that wasn't explicitly given in the problem stem.) (MP8, E, and TEKS-M 1.F)

8. Are there any tools in our classroom that would help us represent the situation (MP5 and TEKS-M 1.C)?

The purpose of this discourse is to scaffold every student's thinking to a higher level to make sense of the problem, think about various ways to begin to work toward a solution, and ultimately find a solution, as well as to give those students who are ready opportunities to stretch their thinking and deepen their mathematical understanding. This routine creates autonomous problem solvers for the future. Students who can make sense of problems can think deeply about how the mathematical relationships are connected to the context and are able to independently work toward a solution for any novel problem.

At this point, students would work on the problem independently for a few days, depending on the complexity of the problem. They could choose to work with a partner, in a small group, or independently, knowing that it will be their responsibility to share their ideas, ask clarifying questions of other students, and determine the precision of their own work and of their classmates.

Finally, the teacher holds a whole-class math forum, or discussion where students share ideas, and partial thinking, engage in productive thinking/arguments about the problem, and ask each other clarifying and challenging questions. This routine creates autonomous problem solvers. After several whole group lessons on the Three Reads problem-solving routine, students begin to autonomously use the routine when new problems are presented. Teachers have found that this simple routine can support all students in solving cognitively demanding and novel problems. As teachers and students begin to speak a common mathematical language, the posed problem begins to make sense and offers students multiple levels of access to the skillset of being a problem solver. Students start to say, "I can do this!" instead of, "I need help!"

Using the Three Reads problem-solving routine supports students in thinking deeply and engaging in discourse at a variety of levels. At one level, it supports a gradual release of responsibility (I do, we do, you do). During the first read, the teacher reads the story aloud, giving students the opportunity to engage with the text through active listening. In this read, the teacher, as facilitator, models that a good problem solver starts by reading to comprehend or make sense of the context of the story (I do).

During the second read, students receive a copy of the problem and read silently as the teacher reads it again to focus on the important mathematical qualities and their relationships (we do).

And finally, during the third read, students read independently and begin to sketch a diagram or picture of the problem (you do). It is a well-accepted routine among educators and can scaffold the autonomous learning necessary to help students problem solve and it can lead up to more engaged discourse in the classroom.

At another level, Daro and Asturias share why this slowing down of the problem-solving process is so important and how discourse plays an important role on their Moving Beyond Answer-getting website:

> **There is nothing wrong with focusing on the solution of a math problem. But students can be observed using truly haphazard approaches to getting a solution. Some teachers refer to this as "mashing numbers." Typically, students try to quickly find the numbers embedded in the word problem and then run through various familiar operations to come up with an answer, regardless of whether that answer makes sense. Rarely do students take time to consider the question being asked of them and the relationship among the quantities in the problem to thoughtfully decide on an approach for a solution. Looking at word problems as three-part entities enables teachers to vary the part or parts that students need to fill in or solve. Students might be given the stem only and asked what questions can be asked with the information provided. Or, they might be given a solution and asked to imagine a situation and question for an answer. Students must then think about an equation in terms of the relationships it represents and apply that understanding to a real-life (or in some cases imagined) situation.**
> **—Daro 2017**

Discourse allows students to share personal thinking, critique the reasoning of other students, engage in productive and respectful arguments about differing ideas, and learn to ask clarifying and challenging questions of themselves and others, producing mathematical learning for all students when solving novel problems.

Conclusion

It can be overwhelming to think about helping all students meet both mathematics content standards and develop a depth of understanding about the actions called for by the mathematical practice and process standards. Yet, these standards beautifully summarize what mathematicians do, how they think, and what kinds of actions they take when solving mathematical problems. Often, many students arrive to classrooms with fixed mindsets about themselves as mathematicians that is often not positive, and our culturally and linguistically diverse (CLD) students feel left out of the conversations.

Today, our expectations for students must go well beyond the ability to memorize math facts and perform basic computational procedures. We need to recognize them as just part of what our students need to know and be able to do. Engaging students in mathematical discourse can be a starting point for meeting standards and building a lifelong curiosity about how to make sense of mathematics.

Let students see your love of math. Let them see you struggle, make mistakes, and take risks. The teaching strategies we choose are the foundation for the beliefs that students form about themselves as mathematicians. Creating a classroom full of mathematical discourse is at the core of growing our community of learners.

Reflect and Discuss

1. Choose one of the habits of mind, and think about how you plan lessons to foster those habits.

2. Think about the questions that you ask students. Which questions can you add to promote deeper thinking and discourse with students?

3. Think about the Four Freedoms to Learning and Positive Norms. Choose two to focus on, and discuss with students as you build a discourse community.

4. What strategies do you already use that support equity in learning mathematics? Which new strategies can you try?

5. Think about the Three Reads problem-solving routine. How is it different from what you already do? How does it promote discourse in a problem-solving based classroom?

Chapter 3

Teacher Moves That Promote Effective Student Discourse

A community approach enhances learning: It helps to advance understanding, expand students' capabilities for investigation, enrich the questions that guide inquiry, and aid students in giving meaning to experiences.

—National Research Council 1996, 46

Learning mathematics is not an isolated or passive activity. It is a process of making sense and establishing meaning, which is built individually and collectively in a community. Helping students discover how to make sense of mathematics doesn't necessarily take longer than directly teaching them how to do the math using procedures and memorized skills; however, it does take a different mindset on the part of the teacher. Teachers have to believe that all students are mathematical thinkers and have the ability to give insightful observations and informed conjectures. They must nurture a classroom environment where students can be themselves and where their ways of knowing, thinking, and expressing themselves are valued.

Thinking about the classroom as a community is not a new idea. John Dewey (1916) is often given credit for creating the idea of schools and classrooms as communities. For Dewey, classroom community was not just another name for a collection of students or an ideal of harmony and collaboration. It was the process of people living, working, and especially learning together. In such a community, the teacher helps students learn with and from each other while respecting diverse ideas, skills, and experiences. Knowing students' social and academic interests and needs is a crucial first step in supporting rich mathematical discourse in any classroom. Weekly planning, daily decisions, and specifically chosen teacher moves are all based on a teacher's level of familiarity with his or her students' mathematical knowledge, background experiences, and learning styles.

So, how do teachers acquire this depth of knowledge of their students to build a community of mathematical learners? First and foremost, teachers must spend time building relationships with students in the classroom and getting to know each student while working toward effective discourse communities. A teacher begins to engage students in productive mathematics discussions by establishing a learning environment that welcomes student involvement and includes expectations for all students to contribute to the discourse community. Teachers will need to develop a vision of the classroom environment with their students from the first day of class. The whole group discusses, negotiates, and adopts this vision, so everyone shares and values it. There are several ideas to keep in mind as you work with your students to create a community of mathematical learners:

- Students participate in establishing classroom behavioral norms in the interest of their mutual well-being.

- Teachers and students consider themselves members of a community of learners.

- Collaborative planning with students helps bring student interests and concerns into the expectations for learning.

- Diversity of viewpoints is valued for enriching learning.

- Teaching and learning are reciprocal processes.

- Attitudes and perceptions are incorporated as elements of the learning process.

It is paramount to take the time to establish classroom community norms and agreements to build a foundation for effective mathematical conversations and discourse. Teachers start this process by sharing personal stories and past learning experiences to help students feel comfortable sharing their own stories. During the agreement-making process, look for indicators that may further enhance the classroom learning community. It is important to explore students' feelings about their past mathematical learning via conversations and journaling. During this time, teachers should emphasize that these community agreements are just as important for the teacher's growth as they are for all students' academic success. Ask students to describe a classroom that can offer positive learning opportunities, and discuss what types of agreements should be in place to create such an environment. Beginning this discussion might sound like: "As a community of learners, we recognize that we can learn more together than by ourselves. For our work in mathematics to be as productive as possible, we commit to follow these agreements. With your table group, list a few agreements the team believes will give optimal learning experiences in our math classroom."

Through the process, have students share personal stories of past learning experiences. Then, ask students to reason about and justify the proposed agreements they want on the list and agree to begin the year attempting to follow the community agreements. See Figure 3.1 for an example of a brainstormed list of student ideas for their classroom agreements. The list may be long at the beginning of the process, but it will be refined, added to, and revisited as the focus on building community continues.

Figure 3.1 Beginning of the Year Brainstormed List by a Third-Grade Class

Room 10's Community Agreement
- Treat others like you want to be treated.
- Listen to the speaker.
- Don't talk out.
- No toys at your desk.
- Think before you do something.
- Follow directions the first time.
- Sit in your seat nicely.
- Participate fully.
- Finish your work on time.
- Raise your hand.
- Keep your hands out of your desk.
- Keep your hands and feet to yourself.
- Help others when they are left behind.
- Do not get a drink when the teacher is teaching.
- Find compromises.
- Pay attention.
- Be polite.
- Do your own work.
- Use wise choices if you have a small problem: It's Your Choice!
- If you have a BIG problem, tell an adult you trust.
- Be patient.
- No talking in the classroom library.
- Allow others to learn at their own pace.
- Try hard.
- Do your best!

For these lists to be true agreements, the teacher records all ideas and uses students' own words. It is important for the teacher to avoid projecting his or her rules during the process. This allows students to take ownership of the community norms. Teachers can suggest agreements to be included in the brainstormed list but must be willing to

Emotional intelligence
is the ability to identify,
assess, and manage the
emotions of the teacher
and students.

compromise and work toward agreements as a community of learners. Often, this early student-generated brainstormed list takes on a negative, rule-oriented air. It looks like a black-and-white, do-or-do-not list. But without it, there is no buy-in.

These beginning rules are often social norms, such as listening to each other, participating fully in an activity, being a risk-taker, using a quiet voice, and using manipulatives as tools not toys. It is the classroom community's job to develop these into sociomathematical norms (Yackel and Cobb 1996). Student-developed sociomathematical norms might include: preparing to start and discuss a task; listening with the intention of understanding and comparing strategies; asking genuine questions of each other for clarification; checking for understanding and encouraging justifications; asking, not telling; providing and honoring think time; keeping each other accountable; and maintaining a safe environment for risk-taking. It is therefore the teacher's job to revisit the list and role-play, modeling what these behaviors look, sound, and feel like when engaged as a community of mathematicians. Agreements should be revisited daily at the beginning and end of each math period. Both the teacher and students should model and role-play appropriate choices and have discussions that help them focus on the norms of being mathematicians.

One area that supports strong community environments happens when teachers focus on student emotional intelligence. Teachers get to know students in a deeper manner and build strong, mutually respectful relationships. Teachers need to show students this level of genuine care. They will see that creating effective discourse and asking students to be risk-takers to share thinking, strategies, and ideas requires a deep level of trust. All students must trust that everybody is on their side and no one will fail or hurt them because everyone genuinely cares about them as learners in their community.

Establish this genuine environment by providing reminders of agreements. This is done by explicitly and transparently modeling high expectations over and over for the class to internalize them and build a trust. Teachers can do so by focusing their energy on observing interactions and listening to individual students, as well as groups. Teachers need to constantly move within the classroom and never sit in one place or spend more than a few minutes in one location. As teachers actively listen to students, they will see the mathematical knowledge students bring to the new learning community. Observing students' mathematical thinking and asking them to self-evaluate, while simultaneously checking for understanding, helps the community of mathematicians grow.

Actively listening, asking genuine questions, and getting excited about student ideas are also critical to this process. The main goal of building community is to emphatically highlight the importance of students sharing ideas and strategies through the use of mathematically sound thoughts, not just words. Teachers who are always looking for these opportunities will begin to confirm how discourse strengthens learning for all. Through very carefully planned and executed moves, teachers will know their students better and begin the process of effectively facilitating mathematical discourse in the classroom.

An Elementary Classroom's Mathematical Community Agreement

Listen to each other and respect his or her ideas, strategies, and thinking.

Ask thoughtful and clarifying questions.

Disagree with others with respect.

Volunteer your ideas in math discussions.

Don't be afraid to struggle with challenging ideas and problems.

Know that it's okay to be confused and make mistakes.

Confusion leads to new learning, and mistakes help clarify our learning.

Have fun and enjoy discovering new things about math!

A Secondary Classroom's Mathematical Community Agreement

As Effective Mathematicians, We Agree To…

Be aware of volume.

Take learning seriously.

Be neat and organized.

Respect others' learning process.

Persevere and never give up.

Be cooperative and work together.

Stay focused and on topic.

Take risks and try new things.

Fight for sensemaking!

Cary Cermak-Rudolf, a third grade teacher, took a student-centered approach. It would be quickly apparent from a glance into her classroom in the middle of the year that she had built and maintained a community of learners. As part of her classroom community, she frequently communicates about her own ability to follow norms. On this particular day, she called on one student, Jen, who had asked to share her strategy with the rest of the class.

Ms. Cermak: Jen has asked very politely to share her strategy. So, please give her your attention.

Jen shared her idea with the class.

Ms. Cermak: Can anyone restate what Jen just said in their own words? I was not following one of our community agreements. I was focused on something else!

It was apparent that Ms. Cermak-Rudolf noticed another student's off-task behavior while Jen shared her idea, and she missed hearing the mathematics. Ms. Cermak-Rudolf was transparent about not following the norm and, by asking another student to restate Jen's idea, she modeled the Stay Focused and On Topic classroom agreement. Ms. Cermak-Rudolf then asked the class which agreement she was working on. This brought the class's attention to the mutually built community agreements. When teachers, such as Ms. Cermak-Rudolf, transparently reflect on their own actions with students about these agreements, it allows them to continually build a learning community where teachers and students know each other well and can encourage everyone to communicate successes and failures in mathematics.

By co-creating community norms, teachers build relationships with students that create a sense of pride in belonging to the classroom community. This sense of belonging sets the beginning stages for the teacher to get to know students. By encouraging students to share stories, past learning experiences, and real-life interests, students learn how to listen carefully, ask genuine questions, and be empathetic with each other. This eventually results in the mathematical discourse becoming more effective.

> We are a team, therefore we talk a lot. We hold class meetings, and we talk about how we're an effective team. At the beginning of the year, we spend a lot of time talking about how everybody has something they are working on, and I share with them that I make mistakes and count on them to support me when that happens, so my brain can grow, just like I want their brains to grow!
>
> —Ms. Cary Cermak-Rudolf

Making Transparent Teacher Decisions

The energy and enthusiasm teachers put into making teaching moves and their decisions transparent throughout lessons support student-to-student discourse. This practice is a way to have students ask themselves about mathematics in any classroom activity, and it promotes autonomous learning. Saying things like, "I think there is a really cool math connection happening here" or, "This problem is really puzzling me right now" outwardly models the teacher's thinking as part of the community. It encourages students to ask questions, puzzle about problems, and reason about mathematics without any coaching.

Another opportunity for transparent talk happens when teachers carefully select and sequence student work to share and discuss with the whole group. Noticing and sharing a common student error, or one made by the teacher, gives students support to take the risk to share their mistakes. It might sound like this scenario started by Diego:

Diego: So, when I was doing this math problem, I noticed that someone else did it the same way that I did. Jill and I think we found our mistake! Can we come up and share our mistake with the class?

Jill and Diego come to the front of the classroom and share their thinking that led to a mistake. They excitedly share what they learned.

Ms. Cermak: If you do not have questions, then we can assume you agree that we are on the right track and our solution makes sense. Does anyone have a challenging or clarifying question for our team?

Here we see Ms. Cermak-Rudolf creating a safe environment for her students to share an error. She gives Diego and Jill the opportunity to learn from their mistakes and turns the learning back to the math community at large. Looking again at Ms. Cermak-Rudolf's teaching style, she uses the phrase "I am doing this because" after a learning opportunity to support students' risk-taking and to challenge them within the safety of the mathematical learning community. This structure encourages everyone to transparently discuss how his or her thinking is alike and different.

To further encourage discourse opportunities in the classroom, use phrases that are open and honest, such as "Your work is not finished yet, but it is the closest I have seen to something we can discuss together." It delivers a not-yet message and builds perseverance. Most students appreciate knowing that their thinking was moving in a positive direction and that it is okay not to have a solution yet. They understand that learning is a process out of which they receive feedback.

Another effective transparent teacher move happens when students talk about the boundary of time. Planning for and setting appropriate time limits throughout a lesson needs to be accurate. Some examples of this transparent move are:

- Now, I am going to set you loose to solve this. You will have 10 minutes to do the problem before we go to _____.

- Hold up your fingers to show how many more minutes you think you need.

- Time is going by so fast! You have two minutes left.

- I am going to give you three minutes. Ready, go!

Using a timer initially to inform this practice is helpful, but the goal is to develop an accurate internal clock, use your best gut feeling, and stick to it. If you change the timeframe, it is important to tell students why you are adding more time and why it is important instructionally. For example:

Thoughts from a Reflective Practitioner

I can no longer imagine not using discourse to teach. Everything I do is purposeful and intentional. Nothing is random. Everyday I reflect on what is working and what is not, and I plan my next day with these things in mind. I think about what is the most effective use of discourse in today's lesson. I want to instill discourse into my learning community.

—Ms. Cermak-Rudolf

Ms. Cermak: I am not going to give you a lot of time to do this because we have a lot of other interesting math ideas to work on today. You have five minutes. Okay, go!

As the teacher wanders and observes her students, she notices that five minutes are up.

Ms. Cermak: I will give you another minute. Some of you are thinking hard and visualizing the shape, *but* in one more minute, you need to determine who on your team can best communicate your group's results.

After a minute, she calls time and begins the discussion.

Ms. Cermak: Now, let's step back. You might be wondering why we are doing this. We have not talked about communication much. Well, we are moving onto some different mathematical ideas. You are great problem solvers. You just showed me that on our state test! But, we are not very good at explaining our thinking aloud, so I gave you more time to figure out how to tell us exactly what your pictures look like. Tell us exactly what you did and why it makes sense to you. You have only three minutes to do this. Ready, go!

After exactly three minutes, Ms. Cermak-Rudolf gives a signal and talks with the class about their ability to find appropriate ways to communicate mathematical ideas to each other. She uses a pre-determined protocol for students to communicate their mathematical ideas and discuss how they talk to each other. This transparency about why they are talking about communication in mathematics is critical for students learning to internalize their need to communicate thinking when solving problems as mathematicians.

The Role of the Teacher's Mathematical Content Knowledge

Teachers will ultimately teach the way they were taught, unless they are instructed differently. Teachers' beliefs about knowing and doing mathematics and about how children come to make sense of mathematics affect how they approach instruction. Having a profound, flexible, and adaptive knowledge of mathematical content is critical (Ma 1999). Many math teachers often purposefully avoid uncomfortable math topics and sometimes dance around the mathematics because of their lack of confidence regarding content knowledge and depth of understanding.

In Deborah Ball's extensive and ongoing research (Ball, Thames, and Phelps 2008), she discusses four types of content knowledge every math teacher needs. The fourth category is labeled Specialized Content Knowledge (SCK). This level of mathematical knowledge is beyond that expected of any well-educated adult. But, it carries great weight. Teachers' SCK enables them to be adaptable for when students become curious. And it is student curiosity which is the prize that can engender new mathematicians and can deepen comprehension.

In this area, there are several mathematical tasks of teaching presented and represented in Figure 3.2 on page 69.

Figure 3.2 Mathematical Tasks of Teaching

Presenting mathematical ideas

Responding to students' "why" questions

Finding an example to make a specific mathematical point

Recognizing what is involved in using a particular representation

Linking representations to underlying ideas and to other representation

Connecting a topic being taught to topics from prior or future years

Explaining mathematical goals and purposes to parents

Appraising and adapting the mathematical content of textbooks

Modifying tasks to be either easier or harder

Evaluating the plausibility of students' claims (often quickly)

Giving or evaluating mathematical explanation

Choosing and developing useable mathematical definitions

Using mathematical notation and language and critiquing its use

Asking productive mathematical questions

Selecting representations for particular purposes

Inspecting equivalencies

Printed with permission from Ball, Thames, and Phelps 2008

Which of these tasks are easy for you? Which are tricky? These are things teachers routinely focus on when teaching mathematics. This rather long and daunting list of requirements makes teaching mathematics complex. It is a knowledge that most mathematicians do not necessarily need to do their jobs. As teachers of math, we need to foster a desire in students and in ourselves to learn specialized knowledge. It will take time and support to develop these specialized teaching skills. Yet, taking the time and asking for support when deepening content knowledge and pedagogy supports productive mathematical discourse within K–12 classrooms.

Decentralization of the Teacher

According to Chapin and Eastman, the learning environment and classroom culture that a teacher establishes has a tremendous influence on students' attitudes toward mathematics: "Constructing learning environments that develop communities of learners is more complex than simply encouraging discourse and asking students to complete mathematical tasks" (1996, 115). Teachers themselves need to be learners, thinkers, and risk takers.

Let's return to Ms. Cermak-Rudolf's third-grade classroom for a deep dive. As you walk into the middle of Ms. Cermak-Rudolf's mathematics lesson, her students are nestled around a task on equivalent fractions. You hear comments from students, "Can you prove that?," "Are you sure it works that way?," and "I don't see it. Can you show me again?" At first glance, there is no teacher in sight. Ms. Cermak-Rudolf is kneeling next to a group of students and carefully listening. At a pivotal moment during group discussions, Ms. Cermak-Rudolf pulls students together for a discussion where she has carefully selected students for presenting their solution strategies and accompanying rationalizations. What follows are bursts of discourse among students that culminates into mathematical understanding. It leaves students with a feeling of satisfaction, joy, and a growing desire to learn more (Higgins, Cermak-Rudolf, and Blanke 2009).

Ms. Cermak-Rudolf's ability to decentralize her role as the facilitator of classroom discourse and turn learning back to students is a critical pedagogical move. Teachers must make a concerted effort to understand the mathematical thinking and perspectives of their students, thereby enabling better student engagement in meaningful discourse.

However, this does not mean learning comes from the front and center of a classroom. When it comes to rearranging where learning takes place, many teachers share how disequilibrium during mathematics lessons challenged them. It is the personal challenges and gradually increasing "Aha!" moments of students that provide self-confidence to feel comfortable with a discourse classroom, but it will take time and risk-taking on the part of the teacher.

When teachers develop a desire to understand and adopt the perspective of their students' thinking, which can often be quite different than their own, an important part of the decentralization of teaching begins. Discourse begins. If this becomes a goal for every lesson and is carefully planned and executed, the teacher has the potential to become a co-learner and facilitator of student-to-student discourse. Teachers can see the importance of "putting themselves in students' shoes" to encourage them to take risks, share ideas, and engage in discourse. All of this helps to continually build an equitable learning community that provides all students an opportunity to thrive.

Teachers might self-reflect by asking the following questions during and after a lesson:

- Why did I stay with that student for so long?
- Why did I move to the side during that discussion?
- Where was I when I asked that question?
- How did I facilitate the discussion by my placement in the classroom?

Many teachers may not know why they did what they did. But by reflecting with these questions, teachers begin to think about where they physically choose to stand or sit and how they move. Asking a coach or principal to watch a lesson and focus on just a teacher's physical placement in the classroom can be enlightening. This kind of reflection helps teachers realize the goal is not to be the authority on how to learn mathematics. It also helps free them so students view the teacher as a co-learner. When posing novel problems or asking cognitively demanding questions, teachers need to distance themselves. This way, students are not sitting and waiting for the answer.

Another reason decentralization of the teacher is important is to avoid the student and class constantly directing their attention to the teacher when a student stands in front of the class to share his or her ideas. When teachers move to the back or side of the room, students can focus their attention on student talk and ideas. Teachers can then become part of the discussion but not be seen as the leader of that discussion. This is a gradual shift that progresses with conscientious practice and movement by the teacher. Decentralizing your physical position in the classroom supports student-to-student discourse.

Another purposeful teacher move that helps students take over discourse is purposeful and consistent visiting of table groups while students work. Teachers who listen, ask genuine questions, and then walk away turn the learning back to students. When teachers do this, it requires students to make connections on their own. At a certain point, students do all the work and, of course, the learning!

The chosen physical arrangement of the classroom is another teacher move that supports successful decentralization of the teacher. Desks arranged in table groups of four to six students support eye contact between students and allow the teacher to easily move around the room. When there is no real front or back of the room, learners focus on each other instead.

It is also important to constantly rotate where and with whom students sit. The more often a teacher physically rearranges student seating, the easier it is for all students to effectively collaborate with each other as partners, table group members, and whole class. Everyone becomes willing to learn with each other and look forward to new partners and table group members. Doing this weekly or monthly will build community and the culture that "we are in this together."

The Role of Teacher Talk and Focusing on the Learner

A teacher's language choice plays an important part of developing a discourse community. When teachers choose language or teacher talk that uses "I" statements, they focus students thinking toward a vision of mathematics that must be like the teacher's. Sometimes, this simple word can send conflicting messages to students regarding their emerging ideas versus what the teacher wants. For example, when teachers share statements like "I did it this way" or "I learned it this way," students may think "this way" is the best way to think about that problem.

Another hindrance to this type of teacher talk is that students might think making a mistake or being in a state of disequilibrium or confusion is not a good thing. As teachers, we must choose our words, feedback, and questions carefully if we are to create a rich discourse community. This has been acknowledged (Ma 1999, Boaler 2016, Herbel-Eisenmann and Cirillo 2009) as one of the big roadblocks to change for teachers who believe they need to tell students how to do things step-by-step.

To begin developing discourse, teachers must often change the type of questions being asked. Research shows that teachers who ask funneling questions versus focusing questions have less student-led discourse in their classrooms. Herbel-Eisenmann and Breyfogle discuss how moving from funneling to focusing questions can be a key teacher move when orchestrating a discourse-rich environment (NCTM 2005). Funneling occurs when the teacher asks a series of questions that guide students through a procedure or to a desired end. With this type of questioning, the teacher engages in cognitive activities, but students are merely answering questions to arrive at solutions without making mathematical connections. This is not good. This type of questioning and teacher talk limits what students can contribute because it directs them to think in a predetermined way based on how the teacher solves the problem. No insight is created, and it restricts the depth of discourse possible in the classroom. When discourse is absent, teachers lose the ability to reveal student understanding or misconceptions of mathematics.

The alternative to funneling questions is focusing questions. Focusing questions require teachers to closely listen to students' responses and guide learning based on what students are thinking rather than how they would solve the problem. It is important for teachers wanting to engage in genuine discourse with their students to avoid funneling questions and to truly engage their students through careful listening and asking focused, genuine questions to develop artful discussions.

Asking Genuine Questions

One of the most commonly used forms of teacher questioning is what is referred to as the Initiate-Response-Evaluate (IRE) pattern, in which the teacher **initiates** a question, the student **responds** (usually with a solution), and the teacher **evaluates** the student response as either right or wrong (Mehan 1979). This type of exchange teaches students to guess or find the answer they believe the teacher is looking for (Sherin 2002; Wood, Cobb, and Yackel 1991). Students then wait for the teacher to be the authority who determines whether their answers are correct. This relationship makes students dependent on others for judging whether their thinking/answer is on target.

Boaler and Brodie (2004) examined hundreds of math lessons and analyzed teacher questions. They found many teachers asked information gathering questions that required immediate answers, rehearsed facts/procedures, and questions that enabled students to state facts or procedures following the IRE pattern. However, in classrooms where teachers used a variety of question patterns and included questions that explored mathematical meanings and relationships, or probed students to explain their thinking, a rich student-led discourse environment and higher mathematical achievement were generated. Teachers who purposely prepared questions by anticipating student responses for instructional tasks were more apt to know when and how to use genuine questions that deeply probed student thinking.

Teachers can engage students in discourse by posing genuine questions to encourage discussion and debate, and to require students to attend to the mathematics at hand while explaining and justifying their thinking. These questions are *genuine* because the teacher does not know the answer to them and is sincerely looking to deepen his or her understanding of student ideas. They are posed so teachers and students can see ways of thinking other than their own. Different questions elicit different types of discourse and learning opportunities. Genuine questions that promote mathematical thinking and discourse can help students engage in the habits of mind of a mathematician like those found in Figure 3.3 on page 74.

Figure 3.3 Genuine Questions for Teacher Talk

- What did you notice about _____?
- What if...? *(Conjecture)*
- Can you prove that?
- How do you know what you know? *(Evidence)*
- What part do you agree or disagree with? Why?
- What do you think?
- Do you see any patterns here?
- What do you predict will happen next?
- How did you figure it out?
- Does anyone have a different way to solve this problem?
- Can you convince us?

To train ourselves to ask genuine questions instead of closed-ended questions (e.g., "How much is 5 × 7?"), it can be helpful to make posters of your top 12 questions and display them in your classroom. Then, purposefully plan which questions to use in every lesson. Although a list of questions is not exhaustive, having some specific questions ready in advance of a lesson means you do not have to develop questions on the fly. Developing questions on the fly can be very difficult for teachers juggling a classroom full of students with different needs and levels of scaffolding. In these moments, it is easy to revert to old questioning habits.

Posting genuine questions in the classroom can help teachers shift student talk as well by asking students to genuinely question each other rather than only sharing solutions to problems. Strategically using open-ended questions focuses learning on the process rather than the solution to engage students in mathematical discourse.

As you refine the art of question planning, you will see how different questions move discourse in different directions. Certain questions elicit different types of student responses and thinking. See Figure 3.4 on page 75 for questions in various categories that produce different learning outcomes. These questions can be found on the Digital Resources (questions.pdf).

Figure 3.4 Questions to Elicit Mathematical Learning

Questions to Elicit Mathematical Learning

Students working together to make sense of mathematics:

- Do you agree with _____'s thinking?
- Do you respectfully have a different idea?
- Can you restate _____'s thinking in your own words?
- Can one of you answer _____'s question?
- Did anyone have a different way of thinking about this problem?

Students relying more on themselves to determine whether something is mathematically correct:

- Is there something you know for sure about this problem?
- Is this problem related to anything else we have been learning about?
- What is your best guess?
- Can you tell me, in your own words, what this problem is about?
- How can you explain what you know right now?

Students learning to reason mathematically:

- Will that always work?
- Can you prove that?
- How is your mathematical idea alike or different from _____'s?
- Will your strategy work for all problems?
- How did you organize your thinking about this problem?

Students learning to make conjecture, invent, and solve problems:

- What would happen if _____? Why?
- Are there any other possible answers?
- Do you see any patterns?
- What do you think comes next?
- How do you know what you know?

Students learning to connect mathematics, its ideas, and its applications:

- How is this problem like something else we have learned?
- Can you write another problem related to this problem?
- What is a similar idea? What is a different idea?
- Will that always work?
- What is the relationship of this strategy to other strategies we have used?

Students learning to comprehend mathematical problems:

- What is the context or situation of this problem?
- Can you tell me what this problem is about in your own words?
- What mathematical quantities are important and how are they related?
- Are there any other important mathematical ideas in this problem?
- What do you already know about _____?

Students focusing on making connections between past and new learning:

- How is this problem similar to something you previously solved?
- What do you know about _____?
- How is that idea mathematically important to this new problem?
- Does anyone have another experience using this type of math?
- What strategies have you been successful with when solving similar problems?

Students seeing the mathematics in activities:

- What was something you learned about _____ today?
- How does today's activity use mathematics?
- What mathematical strategies/skills did you use today?
- What are the mathematical relationships between _____ and _____?
- What mathematical ideas did you find in this problem/activity?

Students persevering when problem solving:

- Can you think of another recording method that might support your thinking?
- Is there another method/strategy you could use to get to the same answer?
- What have you tried so far?
- Is there something you know for sure about this problem?
- Will that always work? How do you know?

Students evaluating and reflecting on their own learning:

- What do you need to do next?
- What would you tell your partner about how you approached this problem?
- How did your ideas contribute to your group's success in solving this problem?
- What was easy for you? What was difficult?
- Was there a point where you had a mathematical "Aha!" in today's lesson?

As you read these questions, it is equally important to acknowledge what they do not do. These questions do not replace the thinking of students by praising student answers or providing too much information, nor do they provide a quick road to finding or confirming answers. These questions are designed to scaffold student thinking, encourage students to reflect, and challenge students to think harder and more deeply about the mathematical ideas at hand.

Encouraging Disequilibrium

When teachers personally experience new learning because they have been put into a state of disequilibrium, they often realize that this state creates deeper mathematical understanding. Jean Piaget (1970a) defined *disequilibrium* as "a conflict between new ideas and current conceptions." Students feel success in mathematics even without mastering a particular concept if they understand that struggle can be an expected and essential part of learning. Susan Carter (2008) found that by having her students identify when they were in disequilibrium, they were able to continue to struggle with a concept or problem without worrying about getting it right. Her findings support the idea that planning for and allowing time for students to work through productive confusion supports long-term learning.

Productive confusion is also established when the teacher and students work in a supportive and collaborative environment. Everyone has a sense of shared purpose in the mathematical learning. This is created when teachers establish warm and supportive relationships with and among their students. Everyone's learning is based on the collective desire to make sense of the mathematical world. When teachers provide concrete experiences, social interactions, and reflection opportunities during math lessons, students often encounter perspectives that contradict their current points of view of learning mathematics. Students who view their struggles with problems and ideas as a constant work in progress learn to reformulate their conceptions about their learning opportunities in mathematics.

This was evident as a third grader, Patrick, candidly shared his thinking about learning math in his journal:

> Although math is sometimes hard, it is still fun too, kind of like a puzzle. I learn many things about explaining my answer, showing proof, and learning new words to help me explain my strategies and thinking even when I am not done. Sometimes I need help and ask my neighbor. And we work it through. Math is exciting because you are not always sure about what you will learn next.

When teachers create moments of disequilibrium for students, instead of directly modeling mathematical ideas, they engage students in productive struggle. They have the potential to see many "Ahas" from their students like Patrick's. If learners do not struggle while learning, they will not own it. Learners have to feel the emotion behind the action. They want to learn how to solve it because they don't want to leave with an uncomfortable feeling. Often, if students are in disequilibrium, they pay more attention to the learning at hand. It makes learning math an emotional investment. Ms. Kimberly Kelly, a middle-school teacher, shared why it is important to plan math lessons that allow time for students to "muck around" with a problem. She said:

> The reason for letting them muck around so long on the problem was that I didn't see anything that was rich mathematically. Students' engagement level is how I decide when I need to stop and when I need to move forward. I knew from past experiences that they needed to grapple with this problem and that what we were about to try to accomplish was going to be really difficult this week. I knew they would need some kind of emotional investment toward trying to make sense of the math. This takes time, and it was well worth the time. Their thinking eventually was astounding!

Often when teachers, such as Ms. Kelly, purposefully make an error, they are hopeful students will catch them. They ask students questions when they (the teachers) already know the answer. This strategy provokes deeper thinking for later student arguments, allowing teachers to check on student progress. The class can then arrive at a tentative agreement about the mathematical learning at hand.

Teachers will sense that working through periods of disequilibrium often leaves students feeling like they have made a great mathematical discovery. It builds self-confidence and student ownership of mathematical understanding until it becomes the norm. It will happen before your very eyes when discourse is carefully facilitated in all lessons.

Promoting Risk-Taking

When mathematics is new or tricky for students, it can be difficult for teachers to encourage student risk-taking. But teachers must respect students' developing thinking process. They can do this by asking permission from students to share their work and deliberately celebrate partial thinking as often as possible.

Teachers might ask students:

- How does this tell the math story?

- How is it like your thinking? How is it different?

- I saw some people do this… Does that mean the same thing as this?

- How can I describe that to you?

- Better yet, is there anyone who sees something different and is willing to share what he or she knows?

Being able to support risk-taking with novel problems or tricky math concepts is important. One of the most critical things teachers need to make time for is solving the mathematics before teaching it. It helps teachers think about students' possible misconceptions, as well as potential successes. When teachers do this, they often realize students are highly capable of solving problems in multiple ways and teachers can anticipate which students may struggle. This gives teachers the opportunity to purposefully plan lessons and choose cognitively demanding problems that allow them to celebrate multiple strategies, including student misconceptions.

The careful planning of discourse promotes student risk-taking and scaffolds learning for all students. Students need to have examples that sharing partial thinking, misconceptions, and correct solutions as genuine learning opportunities can be used to deepen mathematical understanding. Getting students to take these kinds of risks early in the year is initially difficult. Convincing students it is okay to share mistakes with partners and table-group friends is the starting point for being able to do so in front of the class. Learning to share ideas, listen, argue points of view, and ask clarifying questions must begin in a safe environment.

Teachers must take time to explicitly talk to students about the importance of making mistakes as a way to continually learn. Often, teachers do this by sharing their own personal risk-taking experiences in mathematics with their students, as well as other real-life contexts. Through this sharing, teachers communicate their genuine care and honest belief in students' abilities to learn. It gives them a springboard for becoming risk-takers and builds the strength of a true math learning community. It can be coupled with carefully chosen tasks that are cognitively demanding yet have entry points available for all students.

Celebrate risk-taking throughout a lesson and share ideas about what it looks and sounds like often, if not daily. It is imperative to hold class meetings about how to share ideas effectively, ask clarifying and challenging questions, and disagree respectfully with each other. In these meetings, it is important to talk and model what it looks and sounds like to be frustrated, to get that uneasy feeling in your tummy, to be unsure. Getting frustrated or being confused is a good sign and means students are going to learn something really BIG! If students never feel this way at school, they do not need to attend because they already know everything. Therefore, when students feel something is difficult or hard, that means they have a chance to learn something new and think about it in their own way.

As students become risk-takers, celebrate it! For example, when students share strategies or partial thinking and change their minds midstream, right or wrong, it can be celebrated. In these cases, students are also disagreeing with themselves and are learning something important. They are acknowledging the clarifying and challenging questions they are being asked. Often, it is this student-to-student discourse that can get their thinking back on track and help them to brainstorm other ways to solve the problem at hand.

The Role of Time

TIME! TIME! TIME! It just gets in the way in the educational world. There are time issues inside and outside the classroom. Some of them are:

- time to plan
- time to do the mathematics before teaching it
- time to reflect on lessons and teacher moves
- time to reflect on individual and collective student learning
- classroom interruptions, such as fire drills
- teacher time out of the classroom, as well as the many hats all teachers wear to support students as individuals

Realities of being a classroom teacher can hinder effective classroom discourse. Yet, making effective use of time is critical to discourse—every minute counts. Use of protocols, signals, and routines is key. Teachers need to rebound from interruptions and refocus on the importance of timing and pacing to work toward effective mathematical discussions. These pedagogical practices will be discussed on page 81.

Pacing and Timing

Another instructional move that needs to be artfully planned and executed is the pacing of lessons. Many resources and texts are organized into chunks of one-hour blocks of time and usually cover one big mathematical idea over several days. This means teachers must be very aware of the time they spend on each concept, activity, and discussion on a daily basis. Teachers can keep a fast yet appropriate pace within every lesson by setting high expectations for students to complete their thinking in a determined amount of time while the teacher monitors. They can use a timer and time marks in lesson plans to stay on track. When teachers continually talk to students about the time they have to complete tasks, it can assist them in successfully staying within time frames and building efficiency. Some ways of doing this can be seen in the following examples of countdowns:

- "I am going to give you about three minutes to finish discussing, and I need everybody to build consensus. Everyone will have the same answer on each page. Understand? I am looking at the clock. Ready, go."

- "Okay, look at the time."

- "Hold up your fingers for the number of minutes you need to be ready to share as a group. Zero? One? Two?"

When lessons are carefully planned (i.e., using phrases like those above) and executed with pacing in mind, there is no wasted instructional time and student engagement in the learning outcome is high.

Carefully Selecting and Sequencing Student Strategies and Work

As teacher moves are carefully planned and executed, teachers need to make decisions about the direction that discussions will take. It is important to return to the big mathematical ideas central to learning and reflect on what students know and understand related to the mathematical goal. Teachers should select ideas that enhance the mathematical understanding of the class, not just choose a cool idea. They can do so by giving students autonomy to learn the math concepts. But, redistributing opportunity isn't so simple. For example, many teachers use equity sticks (craft sticks with students' names on them) to let students share. When teachers use this strategy, they can lose the ability to control the conversation and efficiently make connections to learning targets or outcomes. They experience random student sharing, which is unproductive for discourse. But, by sequencing ideas, teachers have a tool to refocus random sharing and keep learning on track.

There are different purposes for selecting and sequencing student work during lesson planning. Sometimes teachers look for correct and incorrect solutions and sequence the work to bring out misconceptions as well as effective strategies. It is often helpful to do this by beginning with correct strategies and then moving to commonly observed misconceptions. Teachers may also sequence a set of lessons to see the progression of strategies from concrete models to logical arguments/conjectures to prove a math concept. Another purpose is to see how various models and strategies show the same ideas. All of these purposes connect between student thinking and big mathematical ideas.

Even for the most seasoned teacher, the art of connecting ideas through carefully planned and executed questioning is challenging. Crafting the questions that will make the mathematics visible and understandable is the goal. It takes understanding the mathematics, connecting ideas, and understanding representations and strategies connected to critical mathematical concepts to plan for these questions. All these decisions, by season and novice teachers alike, must be guided by explicit learning goals of the lesson. It is the foundation of every math lesson.

Pedagogical Practices to Engage Students in Mathematical Discourse

To achieve regular discourse in a learning community, there are some basic practices teachers can employ. These practices come from student need. Many come from experts in areas of education that may not have considered a discourse application when they developed them. But, these practices are versatile for any teacher, and especially for the discourse classroom.

Private Think Time

The number one thing students mention when making mathematical community agreements is that they dislike when students share answers while other students are still working on the problem. Teachers and students agree, as soon as a student raises his or her hand with an answer, or the teacher calls on one student to answer, all thinking ceases. This is why private think time is a critical pedagogical practice for teachers. Smith and Stein (2011) assert that "Giving students time to compose their responses signals the value of deliberative thinking, recognizes that deep thinking takes time, and creates a normative environment that respects and rewards both taking time to respond oneself and being patient as others take the time to formulate thoughts" (72). Not allowing students to raise their hands to respond is a way to promote discourse by giving students the gift of time to think.

Use of Protocols

When teachers use protocols, they encourage all students to talk and listen, to diagnose and speculate, and to wonder and generalize about mathematical ideas. Protocols can teach students to be good listeners and notice others' ways of thinking, to pause and think about what they want to say, to work on problems without rushing to answer, and to speak less or more, depending on their normal behavior. Many protocols ask students to question others, suspend judgment, and withhold personal responses at times, as seen in the following teacher protocol:

> **All right, teams, you are going to start by having the person closest to my desk share his or her ideas about the shape that will be made from the net. You will either agree or disagree and ask questions or talk only about that person's work. Does that make sense? Then, the next person shares, and you will follow the same procedure. When your team is done and you all agree that each chosen shape works or doesn't work, you will put your heads down. Clear? Ready, go!**

The use of protocols supports teachers in creating a learning environment where all students (and the teacher) are asked to construct their knowledge and become autonomous learners. Protocols give students equal access to mathematics and enable them to attain deeper understandings of the concepts. Some protocols assign roles to members of a team. These roles have to be well established from the beginning, modeled, role-played, practiced, and revisited throughout the year. Then, a teacher can assign a role, determine a protocol, and get students right to work with no loss of instructional time.

Other scenarios may not be as easy to determine. For example, if you were to explain how to build a three-dimensional shape with straws and clay to a second grader, what might you tell him or her? Let's see how one teacher chose an appropriate protocol:

> I would like the person closest to the TV to pick up the yellow sheet of paper from your table. After I give the directions, you can decide whether you want to keep the paper or give it to someone else in the group. But, if I catch any team saying "Give it to me, give it to me," then the paper goes away. Does that make sense? It is each person's decision. Each person will hand the paper to someone else if he or she doesn't want the job any longer. So, what is going to happen with the yellow paper? The person with the yellow paper is the one in charge of taking notes. He or she is the recorder/reporter. That person does not have to report back to the group. But, that person has to record everything that everyone notices, observes, or any patterns that are seen by the group. That way, you won't forget anything your team noticed. So, if we were to take your packet over to another class, they would know as much as you do. Does that make sense? If you are okay with being the recorder/reporter, then get your pencil out. Or, you can quietly hand it to someone else. Ready?

Within the structure of this protocol, the teacher was able to give students a choice of the role they were to play, which further encouraged autonomous learning. In the next excerpt (see page 84), the teacher uses a protocol to organize group discussions around some carefully selected and sequenced student work.

Ms. Kelly: We are going to look at some of your graphs today. Here is the first one.

Teacher places it under the document camera.

Ms. Kelly: We are just going to think about this one first. Put your thumb up when you have some observations that you notice.

Teacher waits until most thumbs are up.

Ms. Kelly: Oh, excellent! I see that lots of people have made observations. You are going to share in your teams some of the things you noticed. We will start with the person closest to my desk, and they will begin by sharing one or two things. Then, the next person from the team will do the same. What happens when we get back around to you? Can you say something new?

Students all respond YES!

Ms. Kelly: Raise your hand if you are the one who will start. Excellent, let's go!

Students engage in the protocol with the first graph and share ideas as a whole group. The teacher then places the next graph under the document camera.

Ms. Kelly: Use the same protocol where you share one or two things. We will start with the person closest to the door this time. Ready, go!

At its heart, facilitating the use of protocols is about promoting participation, ensuring equity, and building trust in a timely manner. Protocols are deliberately designed with this in mind.

Use of Signals

One of the more difficult aspects of managing student-to-student discourse can be the lack of control a teacher has when bringing students back together to share their thinking as a whole group. Teachers can do this seamlessly with the use of signals. Having a variety of signals available and using them routinely gives students the ability to respond and reconnect when engaged in various types of group work. When these signals are internalized, anyone can use them to get students' attention at any point in any lesson. Simple signals for attention—whether visual (a raised hand) or auditory (a chime, rain stick, or other pleasant-sounding instrument)—are essential classroom management tools. They are more efficient and respectful than calling out, "I need your attention!" Using too much talk to ask for students' attention gives students permission to tune out your voice after the first few words and hinders discourse. It is important to do the following to make signals effective for the promotion of discourse in a math classroom:

- Teach students how to respond to the signal.

- Model exactly how the signal and student response should look and sound, and give students plenty of practice.

- Don't start speaking until everyone is quiet.

- Demand student response and attention! Waiting to speak sends the message that everyone is expected to respond to the signal.

- Be consistent about using signals you've established. Otherwise, signals will lose their power. Students will wonder whether you really mean what you say.

- Don't repeat a signal when it doesn't get students' attention the first time. Repeating signals teaches students they don't have to comply right away. They can wait for the second (or third) repetition.

- Don't expect immediate silence. This may feel disrespectful, but immediate silence is unrealistic. People have a natural need to get to a stopping point in conversations or work. Listen to discussions and make sure students are on topic or on task.

- Use different signal words that complement the math being taught. For example, during a fraction unit, tell students your signal word is *fraction*.

Figure 3.5 on page 86 has verbal and nonverbal signals teachers can implement into their classrooms.

Figure 3.5 Verbal and Nonverbal Signals That Support Discourse

Verbal Signals	Nonverbal Signals
Mathematicians!	Chime signal—Teacher waits.
Eyes on me in 3-2-1. *Eyes on me 1-2-3.*	Timer rings—Teacher waits.
Finish your conversations and focus forward. [Teacher waits and listens.]	Entry task—Work on board as students enter from recess.
Whoever is speaking, finish your sentence and focus forward.	Hand raised—Teacher waits.
Stand up when you are ready to…	Hands on head—Teacher places hand on head says nothing, and waits.
Let's create a signal word for the day. Listen carefully.	Clap pattern—Students repeat clap pattern.
Freeze: Put your hands on your heads so your ideas don't escape!	Music signal—Music begins and fades.

Checking for Understanding

Checking for understanding is a useful pedagogical move in the teaching and learning process and should be implemented often when promoting discourse. The background knowledge students bring into the classroom influences how they understand materials teachers share and lessons, or learning opportunities, provided. Unless teachers check for understanding, it is difficult to know exactly what students are getting out of a lesson. In fact, checking for understanding is part of a formative assessment system in which teachers identify learning goals, provide student feedback, and plan instruction based on students' errors and misconceptions.

Teachers who check for understanding of directions, mathematical knowledge, classroom procedures, and student engagement in every lesson can effectively promote discourse. This "dip-sticking" while facilitating learning and student discourse contributes to the teacher's knowledge of what has been learned and what learning still needs support.

A teacher's constant movement throughout the classroom is one way to monitor student learning. Teachers who casually walk from group to group, listen carefully, stop to ask questions, or make statements to further provoke mathematical conversations are checking for understanding while promoting discourse. As teachers observe a whole-group setting, students can give them various types of signals when concepts make sense, such as thumbs up or other hand signals. These moves tell students it is their responsibility to let the teacher know whether the mathematics makes sense. Teachers often rely on looking at students' eyes, facial expressions, and body language to know whether they are making sense of the mathematics. Instead, students should know it is their job to make sense of their learning and if that is not happening, then they have a responsibility to say something or ask a question. That is what it means to be a lifelong learner.

Teachers can also ask, "How did you know?" or, "Why do you think that?" as a way to check for understanding. These *how* or *why* questions allow room to adjust and modify teaching during the lesson.

Many *what if* questions can also be used to check student understanding and to promote deeper thinking about mathematics. For example:

- "What if there is another way to do this without using numbers?"
- "What if you could do this math in a way that creates a mystery that would trick your neighbor or make them really think?"
- "What if you put another _____ there? How would that change your thinking?"

The ability to formatively assess student understanding throughout math lessons offers teachers many opportunities to promote student-to-student discourse. All students must expect that mathematics has to make sense, and they have to be able to communicate their understanding of the problem at hand.

Motivating Students as Mathematicians

Success breeds success in effective mathematics classrooms. Teachers who can help students be successful and feel satisfied in learning communities motivate students to continually strive to achieve. Creating a fun and relevant learning environment that offers students choices and responsibilities supports academic growth and creates an intrinsic desire to learn more. One of the most important things any teacher can do is take the time to get to know all students well. This begins the process of learning to listen to and reason about each other's thinking. Implementing practices that give students the ability to enjoy learning, no matter where they are in a process, is one teacher move that effectively supports student discourse.

Believing students can surpass their present level of mathematical knowledge is the first step toward motivating them. Make sure students know it is everyone's job to assist each other in the learning process. This will motivate them to learn with and from each other. When high outcomes for all students are an expectation, teachers must hold themselves to those same high expectations. Transparently talking with students about having the same high expectations for the teacher as well as students makes the teacher an integral part of the learning process and further motivates the learning community to achieve sound mathematical knowledge.

Teachers promote autonomous, motivated learning through a combination of moves in the classroom. For example, when students challenge each other by critiquing or questioning, the teacher turns learning back to students and asks them to converse with each other. All students instantly are motivated in discourse about what works and what issues might need to be addressed, while simultaneously providing feedback to the mathematics of the lesson and to each other's ideas.

Teachers can also introduce and enforce routines to motivate students. A lack of teacher presence in student-led daily routines tells students, "I know you can engage mathematically without me." Teachers who enter the room full of energy and high expectations show students they are ready to learn alongside students. This can be a motivating approach.

Often, students are motivated to learn when they encounter real-life experiences that are connected to new mathematical content and learning processes. This can be motivating for students because they see how the learning applies to what's happening in their world. It is important, therefore, for teachers to communicate the joys and struggles in their own mathematical learning to motivate students to share. Making comments like, "I am working hard at understanding this, too, so let's persevere together;" "Do the best you can;" "That's okay. I did not see that happening, but that would be a really good discussion;" "Wow, a great learning opportunity is coming for all of us!" assists everyone as co-learners in the classroom.

When students are motivated to learn, they actively listen and express their ideas orally or in written form in a coherent manner with anticipation for feedback from peer learners who may or not be the teacher.

Being a Reflective Practitioner

Teachers make thousands of snap decisions every day. Our intuitive judgment is developed by experience, training, knowledge, and repetition. Our brains collect information and commit that information to memory to be drawn upon at a later time. Eventually, teachers get to the point where teacher moves and pedagogical processes get pushed down into the subconscious. They become subroutines and are internalized. Reflective practitioners in the classroom have developed natural habits of reflection so they are able to think without thinking. That is, they can make snap decisions about teaching and students. Often, it looks as though they do it on the fly, but it is actually a form of careful planning over time that allows the practice to become automatic. As teachers become reflective practitioners, our lessons shift from the science of teaching to the art of ensuring learning for all students. Teaching ourselves to become reflective practitioners is a first step to effectively engage with our students in a discourse community. The greatest teachers are artists as well!

Conclusion

By establishing learning communities early in the classroom, students can engage in a discourse community that will frame their habits of mind to be mathematicians. It is important, therefore, for students to actively take part in the learning community. They should feel safe to engage and take risks with mathematics. Teachers can support this goal by acting as facilitators in their classrooms. Teachers can assist students in brainstorming community agreements. This will establish students as the owners of the classroom and discussions that will follow. Students should be motivated to ask questions and engage in mathematics with each other. It is by engaging with each other in conversation about strategies and real-world applications that discourse becomes most effective.

Reflect and Discuss

1. Brainstorm a list of agreements you want students to consider as you build a discourse community. How will you begin this conversation with your students?

2. How can you promote productive, healthy risk-taking and encourage disequilibrium among your student learning community?

3. Do you know where you are in the classroom while facilitating discourse? Invite a colleague to observe your lesson and give you feedback about this critical plan.

4. What can you do or say to motivate students to engage in discourse? What questions will you plan to scaffold student thinking and challenge students to reflect upon their ideas?

Chapter 4

How Math Talks Promote Discourse: Arguments, Ideas, and Questions

Many teachers have embraced Math Talks, a brief daily practice when students mentally solve mathematical problems and orally share and reason about each other's thinking as a type of discourse. Math Talks can radically alter teaching, as well as deepen the mathematical understanding of all students in their mathematics classrooms. Teachers can see something magical happen when students learn with Math Talks. These talks become powerful when used successfully because they place the ownership of mathematical learning into the hands of the student community. Students are given permission to make sense of mathematics in their own way, make mathematically convincing arguments, and critique and challenge ideas from their peers. Engaging in discourse can genuinely be seen as fun by students and often fosters a deep affection for the beauty of mathematics. In the same way, mathematicians see discourse among students as central to the meaningful learning of mathematics. As teachers, it is important to carefully prepare and purposefully facilitate student talk, so Math Talks scaffold learning for all students. When teachers commit to teaching mathematics as a sensemaking subject, classroom discourse and math discussions ensue. Many classroom teachers have opted to facilitate Math Talks to achieve this desired goal.

What we often see is the teacher poses a straightforward problem or question, and then thinking begins. The teacher asks students genuine questions, such as "What did you notice?," "Does that make sense?," "Does anyone else have a different way of thinking about that?," and "What is similar or different about these approaches?" Then, students start to eagerly share and constructively argue and support their ideas about why their strategies work. All of this occurs while students learn to enjoy the productive struggle that produces depth of understanding.

Most students love the simplicity and engagement Math Talks offer. Yet, many teachers never experienced a mathematics classroom full of talk, nor were they taught mathematics in this way. Unfortunately, students and parents remember learning mathematics the way the movie *Ferris Bueller's Day Off* depicts class time. Here we see the teacher dryly reading from the board in an onslaught of low-impact discourse: "Anyone…anyone…anyone?" It's no wonder many students dislike school. They are not given the chance to be active members of the discourse community!

This way of teaching often produces boredom and a strong disdain for learning among students. We see the same debilitating behaviors from students in many mathematics classrooms. Students' heads are down on their desks; some students are acting out because they can't make sense of the mathematics; and some students simply never make eye contact with the teacher, hoping not to be called upon. This scenario has proven over the years to be non-productive (NCTM 2014).

Over and over again, teachers pressure students to learn procedures and arithmetic rules to get correct answers when randomly called upon or on a test at the expense of making sense of the mathematics. It takes the joy out of teaching for the teacher and the challenge of puzzling with and making sense of the mathematics for students. When teachers facilitate Math Talks and students engage in mathematical discourse, it can bring joy back into teaching, and excitement for learning back to students. It can also help solve many of the misconceptions students have about solving problems in mathematics.

So many students think there is only one way to get to an answer and that one way is the procedure or algorithm they were taught to use (i.e., rote memorization). Math Talks actually encourage students and teachers to explore their own ideas through discourse rather than worrying about remembering rote procedures. This guides students toward a depth of understanding that emphasizes sensemaking and often offers alternative ways to think about the mathematics. For example, one day during a second grade Math Talk, Mr. Burke was challenging his students to solve $261 - 147$. As students thought about the problem and tried strategies on their whiteboards, he noticed Marshall had written 114. As students continued to strategize, he privately asked Marshall to justify his idea.

Marshall: I used negative numbers in my head.

This piqued Mr. Burke's curiosity.

Mr. Burke: How do you know it is 114? Can you show us?

Marshall proceeded to write the following, with the first three words emphasized.

Marshall: <u>In my head,</u> I used negative numbers: $1 - 7 = -6$. $60 - 40 = 20$. $200 - 100 = 100$. I subtracted 6 from 20 to get 14 because that was an easy fact. Then, I subtracted $200 - 100$ to get 100. $100 + 14$ equals 114!

Mr. Burke was in awe. After Mr. Burke had Marshall share the place value or expanded notation strategy with regrouping ($200 - 100 + 50 - 40 + 11 - 7 = 114$) with the class. Other students asked Marshall clarifying questions to make sense of his strategy. Student-to-student discourse was in full action! Nowhere in the textbook did his strategy appear, nor had Mr. Burke ever seen a student, or adult for that matter, subtract in this manner. Mr. Burke's rather simple problem presented in his 10-minute Math Talk provided a discourse-rich experience for himself, as well as all of his other students.

This example is not uncommon. Both teachers' and students' thinking and learning were stretched and deepened in a significant way. Both could clearly see what they did or did not understand, and everyone asked genuine questions they wanted answered. The entire class monitored and adjusted to the learning in an efficient and expedient manner. In this way, they shared ideas, partial thinking, and strategies. Their Math Talk was a huge success.

Math Talks support a high level of discourse by promoting social responsibility and boosting retention within the mathematical community. Students learn to respect other students' ideas, ask clarifying questions, and challenge each other's thinking. They learn it takes time to understand somebody else's reasoning and misconceptions, or mistakes, as valuable tools in deepening sensemaking of mathematical concepts. They learn it is important to be patient with themselves, as well as with others, and to participate in cooperative conversations as the learning process goes on within the classroom community. Students engage in social aspects of speaking and listening that help them remember mathematical content and justify their arguments.

Let's look at this next scenario: If the teacher makes a statement or directly teaches, some students remember it and some do not. If a student makes a claim and another student respectfully disagrees or agrees with it, the learning process becomes memorable because of the social interaction. Math Talks give the responsibility of the learning to students, which is the overall goal of mathematical discourse and saves students from the unintelligible droning of that economics teacher that is so often the reality in math classrooms.

Another benefit of using mathematical conversations, or Math Talks, is how they support receptive language function development for students. The more we talk in mathematics classes, the greater the opportunity for students to gain a richer sense of mathematical words and phrases, as well as when and how to use them. For students who are emerging bilinguals, this is particularly helpful. It emphasizes that mathematical vocabulary, although explicit, must be understood in various formats for students to comprehend the meaning of a problem. These formats can include spoken language, written texts, diagrams, drawings, tables, graphs, and mathematical expressions or equations. This is a complex task that can be experienced through discourse that happens during Math Talks.

Dr. Hollie (2015) contends that for classrooms to be culturally responsive, teachers need to use response and discussion protocols that "explicitly utilize defined structures for how and when students are to respond to questions and how they are to conduct discussions" (45). This supports all students in learning content in a deep manner by encouraging them to ask questions, build arguments, and share ideas that can challenge and clarify mathematical thinking.

In their book *Strategies for Connecting Content and Language for English Language Learners in Mathematics,* Mora-Flores and Machado (2015) state that when "providing students with multiple opportunities to engage in conversations with diverse partners across the curriculum," we can support both their content and language development (15). It leads us to the idea that mathematics is a language-rich but not language-dependent content area, and students can successfully achieve in a Math Talk environment.

But, learning to reason through discourse takes time. It takes focus on making connections between ideas to generate new thinking and asking robust questions. And both teachers and students need to take time to logically think about their ideas and strategies to build persuasive arguments. To practice this, students and teachers need to engage frequently in discourse to learn how to use talk in strategic ways. There are many different classroom routines that promote Math Talks and discourse. In classrooms, a routine is a whole-class structured activity, which gives students an opportunity to develop mathematical skills over time.

Math Talk routines provide time to

- preview or frontload new concepts;
- review concepts that have been previously explored; and
- practice those concepts, which continue to be fragile or need to be connected to new upcoming learning.

The mathematical focus of a Math Talk routine is the discourse about ideas and building of arguments through questioning and conversation. These talks can be connected to concepts of the daily lesson, or they can frontload or review previously learned math concepts. A good routine lasts approximately 10 to 15 minutes and happens four to five days a week. They can be connected to a mathematical lesson or be part of a distributed practice daily routine, such as calendar or classroom meetings. The key is that students are thinking, talking, and sharing mathematical ideas. No pencil or paper is required. But for a Math Talk to become a routine, it must be introduced and practiced early and often so students can become comfortable with the format and the mathematics becomes more efficient over time.

Why Are Math Talk Routines Important?

The simple answer to why Math Talks are important is because they provide a daily, easy routine that promotes discourse and can be done any time during the day. A teacher can hold a Math Talk first thing in the morning as part of a classroom meeting, after recess or lunch to reset student focus, the start or end of a math session, or even in the middle of a long math block as a transition. Many important math activities require a lot of planning and an hour or more of teaching time, but Math Talks do not. In a fairly short period of time (10–15 minutes), teachers can offer students great mathematical opportunities to talk about ideas, develop arguments, and ask clarifying questions. Math

Talks can support students' evolving views of what mathematics is, develop number sense, help students see the value of mental math skills, and develop the habits of mind mathematicians use when engaging in creative puzzling, and open mathematical problems.

Math Talks encompass many differing routines that have been incorporated into math classrooms for years. They can incorporate problem strings, number study, incredible equations, problems of the week, ratio tables, and number talks. They can also provide a variety of tools, such as solving problems, modeling, and visual models.

Number talks were coined in the early 1990s by Ruth Parker and Kathy Richardson and have been further developed for the secondary classroom by Jo Boaler and Cathy Humphreys (2005). Cathy Humphreys and Ruth Parker (2015) then collaborated to continue the conversation about *Making Number Talks Matter* in their recent publication to develop mathematical practices and deepen student understanding in the Common Core era. Sherry Parrish (2010) was deeply influenced by these pioneers' discourse work and has written a practical K–5 classroom resource with an emphasis on number sense and mental math, along with Ann Dominick. Parrish and Dominick added a number talk resource for grades 3–8 to the extensive collection of resources. It centered around fractions, decimals, and percentages that emphasize discourse as a tool to deepen students' mathematical thinking. Another major contributor to the idea of Math Talks is Pamela Weber Harris (2011, 2014) who, through the use of problem strings in the middle and high school classroom, helps build powerful numeracy for all. All of these resources support the use of discourse in K–12 classrooms. Because Math Talks focus on discourse, it is good to explore mathematical concepts and skills beyond number sense and operations (e.g., geometry, measurement, and algebraic thinking).

Math Talks provide daily opportunities for students to grapple with number relationships, make mathematical connections, analyze their justifications and explanations, and communicate effectively to solidify critical mathematical understanding. There are several lofty goals for teachers to keep in mind when conducting Math Talks.

Math Talks can achieve the following:

1. Develop mathematical fluency in basic computations and procedural skills with an underlying understanding of mathematical concepts;

2. Develop students' ability to recognize and solve routine problems readily and to search for connections to prior learning to reach a solution when no routine path is apparent;

3. Teach students how to communicate precisely about quantities, logical relationships, and unknown values through the use of sketches, diagrams, models, graphs, symbols, signs, mathematical terms, expressions, and equations;

4. Assist students in making connections among mathematical ideas and concepts, as well as thinking about the use of these ideas in other disciplines and the world around them;

5. Develop an appreciation for the beauty and power of the mathematical world;

6. Help students apply mathematics to everyday life and develop an interest in the wide array of mathematically related career choices; and

7. Provide valuable time for all students to practice and develop mathematical fluency.

Because flexible thinking builds fluency in Math Talks, it is important to remember what fluency includes. Susan Jo Russell (2000) extensively discusses how parents, teachers, and students alike equate fluency with quickly getting to a correct solution. However, in her seminal research, she shares the three ideas that support mental math and computational fluency—accuracy, efficiency, and flexibility.

Accuracy includes careful recording, double-checking results, and knowledge of number facts and relationships. Efficiency implies students do not get bogged down in too many steps or lose track of the logic of a strategy. An efficient strategy is one the student can carry out easily. What is efficient for one student may not be efficient for another. Likewise, what is efficient for the teacher may not be efficient for all students. Flexibility requires knowledge of more than one way to solve a particular kind of problem. Students need to be flexible to choose an appropriate strategy for the problem at hand. Students also need to be able to use one method to solve a problem and another method to double-check results.

Fluency is not simply memorizing facts. It is the ability to easily decompose and compose quantities, shapes, and expressions in various ways while constantly searching for patterns and relationships. It includes a depth of understanding of the meaning of operations and a thorough understanding of the base-ten number system, how the numbers are structured in this system, and how the place value system of numbers behaves in different operations, as well as their relationship to each other and other mathematical concepts. Daily, ongoing routines, such as Math Talks, offer students frequent opportunities to develop true fluency.

John Van de Walle (2010) has talked about what makes quality practice for decades. He states that Math Talks fall into the category of quality practice for students. He defines *practice* as "a variety of rich, problem-based tasks or experiences spread over numerous class periods, each addressing the same basic idea." When engaged in this type of practice, students become comfortable and flexible with mathematical ideas that are emphasized and focused on fluency. What quality practice is not, is the mindless drill of computational procedures that rely on rote memory with little understanding. Lengthy drills of algorithmic skills tend to diminish flexibility and reflective thought in blooming mathematicians and hinder the development of the habits of mind mathematicians thrive on.

When teachers prepare students for and engage them in Math Talks, they create routines that nurture quality practice. These daily routines provide an opportunity for all students to become fluent in the use of effective problem-solving strategies. They provide an increased opportunity for students to develop conceptual ideas and learn to elaborate on them and search for useful connections. Math Talks give students not just a limited opportunity but a greater chance for students to understand mathematics with alternative, flexible strategies. And, most importantly, Math Talks give a clear message about what mathematical discourse really is—a way to discuss ideas, ask questions, engage in argumentation, and puzzle about the mathematics to make sense of it. Math Talks are an excellent opportunity to build ongoing rich practice into any instructional program.

Planning Teacher Talk Moves to Promote Discourse during Math Talks

It has been shown that how teachers respond to and manage students' ideas is one of the most critical aspects for having a discourse-rich mathematical classroom, known as teacher talk moves (Chapin, O'Connor, and Anderson 2009; Humphreys and Parker 2015; Hollie 2015; Mora-Flores and Machado 2015). Below you will find teacher talk moves and classroom protocols that will allow all students, including native speakers and emergent bilinguals, to have equal opportunities to participate in mathematical discourse from the earliest of years.

Revoicing

The teacher or student revoices a student's contribution and often asks for verification of his or her interpretation of the student's ideas. Example: "So, you are saying that…?" This talk move allows the teacher to check in with a student about whether what the student said was heard and interpreted correctly by the teacher or another student.

Repeating

The teacher or student restates another student's idea or contribution in his or her own words and reflects on how it is alike or different from his or her idea to build a community of active listeners. Example: "Can you repeat what he just said in your own words?" This talk move provides another phrasing of someone else's reasoning for students to think about and prompts engagement in discourse. Repeating can be used as a formative assessment of student understanding for the teacher as well.

Reasoning

Ask students to apply their own reasoning about someone else's idea or argument. Example: "What do you think about that?," "Do you agree or respectfully have another idea? Why?," and "How is your thinking different than his?" This talk move encourages students to think beyond their own personal ideas they want to share and focuses their attention on what their peers are saying. It also helps students build connections between differing ideas.

Adding On

The teacher prompts students to add onto another student's idea, or the teacher adds onto a student's idea to increase opportunities for participation from a variety of students. Example: "Would someone like to add on?" This talk move helps elicit more discussion when few students are talking, especially when they are not accustomed to explaining their thinking. It gives students opportunities to focus on what others are saying so that they are able to expand on the differing perspectives within the discourse. This helps get multiple solutions/ideas on the table and can assist students toward a deeper level of mathematical thinking and conversation.

Wait Time

Wait time has been emphasized as an excellent teacher move for decades. The current supporting research on effective discourse discussed wait time as perhaps the most valuable talk move (Blanke 2009; Chapin, O'Connor, and Anderson 2009). Often, teachers are too quick to answer their own questions when no one immediately responds. Students quickly become accustomed to this practice and learn not to think about the question. When a teacher waits, the move provides think time and sets the expectation that students should talk about mathematical ideas and that the teacher will patiently wait for their conversation, argument, or idea. Wait time also provides English language learners, or anyone for that matter, extra time to brainstorm. This time is vital for students to process the questions and formulate ideas, arguments, questions, or thoughts.

Turn and Talk

The teacher poses a question or problem and asks students to think about it. Then, he or she signals students to turn to a partner or small group and talk about their thinking, giving students the opportunity to engage in student-to-student discourse in a safe environment before sharing ideas with the whole group. Turn and talk sounds simple, but the routine needs to be practiced. Routines like this allow all students to share their ideas before class discussions. It gives the teacher the opportunity to listen in and formatively assess what students are discussing. This routine also provides time for the teacher to carefully select and sequence student ideas rather than randomly calling on students.

Agree or Respectfully Have a Different Idea (Thumbs Up, Thumbs Sideways)

The teacher or student asks a question or shares a strategy. They then ask learners to share their thoughts about their thinking by putting their thumbs up to indicate agreement or sideways to respectfully show they have a differing point of view.

Pass It On

Students do not raise their hands but are encouraged to call on a variety of people in the classroom to ask clarifying or challenging questions. A small beanbag or soft item can be tossed or passed to students for all to share their ideas. The beanbag will be continually passed so students know to always be ready. This stresses a sense of immediacy and spontaneity and supports student accountability for participation.

Think-Alouds

The Think-Aloud strategy allows the teacher to model how a good mathematician thinks and talks about problems that are posed. The process is fairly simple and supports all students as they develop the habits of mind of being a mathematician. The teacher begins to read a problem and stops periodically to make predictions, clarify meaning, discuss academic vocabulary, make personal connections, bring in prior knowledge, ask questions, and summarize the problem. This explicit modeling of how to read a math problem can benefit all students as they strive for a deeper understanding of the math problem.

Sentence Frames

Using sentence frames for English language learners during Math Talks serves a variety of purposes. Sentence frames provide the support English language learners need to fully participate in math discussions, to contextualize and bring meaning to vocabulary, to help use the vocabulary in grammatically correct and complete sentences, to have a structure for practicing and extending English language skills, and to have differentiated instruction. Sentence frames are beneficial for English language learners because they can be made for different levels of language proficiency. Sentence frames can also help students contextualize the academic vocabulary necessary to make sense of the mathematics and support language in a variety of activities.

During games, sentence frames can focus students on mathematical vocabulary and create conversation between players without the teacher being involved. They can also scaffold student thinking and provide challenges or support when students work autonomously. Sentence frames give English language learners the support they need to fully participate in Math Talks, using language to develop their understanding of the math content. Figure 4.1 on page 102 has an example.

Figure 4.1 Sentence Frames with Numbers

I rolled _____ and you rolled _____.
    (Number)   (Number)

_____ × _____ equals _____.
  (Number)   (Number)   (Number)

_____ is our target number!
  (Number)

Now, we rolled _____ and_____ and _____
                      (Number)   (Number)   (Number)

_____ and _____.
  (Number)   (Number)

My equation is _____.

My score is _____ and your score is _____.
                  (Number)   (Number)

_____ _____ _____ so _____ win!
  (Number)   (< = >)   (Number)   (You/I)

A more challenging sentence frame could be:

My strategy for solving this problem was _____.

www.mathlearningcenter.org

Printed with permission from the Math Learning Center

Orchestrating a Math Talk Community

Mathematical discourse should be an integral part of teaching and learning, but the logistics of designing, facilitating, and assessing discourse are often overwhelming for teachers. Teacher talk moves and protocols can organize discourse and encourage strategic connections between specific cultural behaviors and discussion designs. Discourse offers social and collaborative opportunities for students to express themselves and work together toward shared objectives (Hollie 2015).

When setting up a discourse-rich environment, teachers have to make decisions as they help students build powerful understanding and come to believe in themselves as mathematical thinkers. The teacher must consider building both the physical and emotional environment that promotes risk-taking by all students. Asking students to engage in conversation requires protocols and routines to be in place. For example, a teacher for elementary school may sit on the floor with students in a circle or semi-circle to model his or her part in belonging within the community of learners.

It is important for students in this environment to have the ability to easily turn to face each other, rather than face the teacher. In a discourse-rich secondary classroom, the teacher is often difficult to find as he or she positions himself or herself to be a member of the learning community rather than the leader. If the teacher is at the front of the room, he or she can record student thinking and ask genuine questions to scaffold student learning.

Careful consideration for the placement of charts, visual aids, document cameras, and related vocabulary resources is important to plan for when supporting levels of talk. It is critical for all students to access needed visuals, including vocabulary word walls or charts, various visual models, and student-developed strategies in the form of anchor charts to fully participate in Math Talks.

For rich, mathematical discussions to occur, classrooms must be safe environments where students want to learn and think deeply about mathematics. The value of all the stakeholders needs to be on learning new things, challenging each other, and working together to solve problems rather than focusing on being the first one to get "the" answer. It is important to teach students about the expectations of talking in math class.

All students need to understand *why* it is important to talk about their own and other's reasoning in math class as well as *how* to learn this way. To begin, the community of learners must set expectations and agreements about how math class will look and feel that year. Some discussion points to help set norms before Math Talks are:

1. Explain that everyone is going to talk in math class this year. Students should expect that everyone will be prepared to talk and be willing to take risks.
2. Ask why it's important to talk in math class.
   - Share that contribution of thoughts or partial ideas are expected of all.
   - Explain that the class can learn from sharing ideas.
   - Explain that the class can learn by listening to other students' ideas.

Web Resources for Math Talks

The following resources for Math Talks promote discourse in your math classroom:

- www.mathtalks.net
- www.visualpatterns.org
- www.insidemathematics.org/classroom-videos/number-talks
- www.sfusdmath.org/math-talks-resources.html
- davidwees.com/content/math-talk
- mathteachingstrategies.wordpress.com/2008/11/24/mathematical-routines/
- svmimac.org/images/Cristo_Rey_-_Middle_Level_Bank.pdf
- https://guidedmath.wordpress.com/tag/daily-math-routines/

- Explain that the class can figure out what can be understood and what might still be unclear.

- Talk with other students to better prepare for using mathematics outside the classroom.

3. Explain that the class is going to share ideas, reason, and debate/argue about various ideas, and that is good.

- Discuss and model the various teacher talk moves (listed above), such as repeating, restating, adding on, giving ample wait time, turn and talk, and reasoning, and why they are important.

- Discuss that all good mathematicians talk about each other's ideas, debate, clarify, ask questions, and argue their point of view to better understand the mathematics.

4. Ask what Math Talk time should look like. Establish expectations of each other as fellow mathematicians. Set community agreements, such as those in Figure 4.2 on page 105, for our mathematical community.

- Make a chart of what Math Talks should *Look Like, Sound Like, Feel Like.* Add ideas to the chart before and after Math Talks.

- Share how everyone will hold each other accountable for these agreements. Explicitly plan for and model the agreements. Examples: "It sounds like you both have a differing point of view. I wonder if you can put both of your ideas together to come up with a solution." Or, "When Ellie is sharing her thinking, I want all of you to be thinking… How is her way similar to or different from your way?"

5. Ask what goals should be set for Math Talk time.

- Explain that during Math Talks, students should be reflective, thinking mathematicians. Students should self-evaluate learning and always be setting new goals. Communication should be open and students should be willing to learn from each other and take risks. The more we share our strengths and weaknesses, the more we can take risks and learn.

Figure 4.2 Community Agreement Posters

As Effective Mathematicians,
We Agree to...
1) Be aware of volume
2) Take learning seriously
3) Be neat and organized
4) Respect others learning process
5) Persevere and never give up
6) Be cooperative and work together
7) Stay focused and on topic
8) Take risks and try new things
9) Fight for sensemaking

Workplaces
Look Like   Sound Like   Feel Like

• Calm
• Students using tools
• Working together

• Whispers
• Talk about math

Goal

• Learning
• Brain is working

A community of...
Learners
• Share materials.
• Listen carefully.
• Clean up.
Help others in the group!

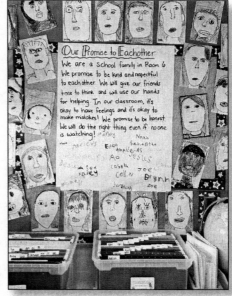

Our Promise to Eachother
We are a school family in Room 6.
We promise to be kind and respectful
to each other. We will give our friends
time to think and will use our hands
for helping. In our classroom, it's
okay to have feelings and it's okay to
make mistakes! We promise to be honest.
We will do the right thing even if no one
is watching!

Courtesy of Jessica Djuric and Kimberly Kelly 2017

Planning a Math Talk

Once the class has created agreements around what mathematical communities will look, sound, and feel like, students will be ready to embark on their first Math Talk. As communities are established, it is important to remember teachers do not teach specific strategies to students during Math Talks. The emphasis is on student-to-student discourse, not on teacher-to-student discourse. Students will learn to think for themselves and determine the most applicable mathematics to the problems at hand.

Before the Math Talk, the teacher must prepare a purposeful mathematical prompt or question. High-cognitive demand questions are best and can come from quality resources or the curriculum. Make sure you start with a good question/prompt. Teachers might ask and answer the following questions as they plan:

- What mathematics will be discussed and emphasized? Why was this problem chosen for the Math Talk (e.g., conceptual understanding; computational strategies; problem-solving strategies; mathematical reasoning; mathematical terminology, symbols, and definitions; mathematical practices; mathematical focus/standard)?

- Will the Math Talk reinforce, refine, and/or extend prior knowledge?

- What intentional choices or values were chosen to scaffold learning?

- What misconceptions and common errors might occur?

Consider how you will respond to common errors/misconceptions by asking:

- What genuine questions will you have prepared to scaffold the discourse and conversation?

- What responses will you hope to hear that will lead to depth of learning for all students?

During a Math Talk teachers will introduce expectations and the mathematical question or prompt. It cannot be emphasized enough that during this time it is the student who leads discussion and the teacher who facilitates. A facilitator during a Math Talk should:

- Elicit and record student thinking;

- Always be inquisitive, not evaluative; and

- Avoid funneling questions to get to answers. Instead, use focusing questions that prompt genuine inquiry and support student risk-taking, arguments, and clarifying questions.

As students break apart the question or prompt, have them share strategies, partial thinking, justification, and proof. This will elicit deeper knowledge and understanding of the mathematical concepts, giving students ownership of the learning process.

Closing a Math Talk has two main goals: remind students of mathematical connections and give students voice. This last part is especially important because it strengthens the community learning environment where students take risks and explore. Have students close the session with what they know and what they are still wondering.

Reflection during Math Talks

For years, the idea of being a reflective practitioner has been discussed in the educational community. Schon (1987) defines *reflection* as "knowing-in-action." In fact, teachers are encouraged to reflect in the moment of action (teaching) in the same way students are invited to reflect on their learning. From this perspective, situations for reflection do not present themselves as givens but are constructed from events that are puzzling, troubling, and uncertain (Schon 1987). Over time, reflection in action is developed and contributes to the rich discourse in Math Talks. Using daily Math Talk reflections, teachers and students can become reflective practitioners, which has proven to be one of the most effective ways to scaffold learning and academically grow (Blanke 2009).

Understand that Math Talks are not really about how many ways problems are solved. They are about how students make sense of the problems. They are never trying to figure out what the teacher wants them to do. Instead, they are deeply engaged in their own sensemaking process. Math Talks and Math Talk reflections provide time for all students to learn they can figure things out on their own. They are autonomous, reflective, and lifelong learners. Teacher and student reflection about Math Talks can make student learning more meaningful.

Before the Math Talk Chart

It is important for students to have an opportunity to reflect on learning. Before a Math Talk, distribute copies of Figure 4.3 found on the Digital Resources (beforemt.pdf). This reflection time should be used for students to make predictions and observations about the math concepts. If students are reluctant to share, encourage them to write what makes them curious and to use math strategy posters around the room, making sure that big ideas come from the student, not the teacher.

Figure 4.3 Before the Math Talk Chart

Before the Math Talks	
What We Know Now For Sure!	**Our Wonderings...**

During the Math Talk Plan

Teachers can anticipate possible issues and solutions that could arise during the session. Use copies of Figure 4.4 found on the Digital Resources (duringmt.pdf) to prepare for a Math Talk. These plans will provide structure to make doing Math Talks easier as more Math Talks are implemented and eventually take less time.

Figure 4.4 During the Math Talk Plan

<table>
<tr><td colspan="2" align="center">**During the Math Talk**</td></tr>
<tr>
<td>My Math Talk Plan

Grade Level: _____

Whole Group   Small Group

Date: _____

Time: _____</td>
<td>Big Math Idea:

Prompt/Problem:

Source:</td>
</tr>
<tr>
<td>Do the Math! Anticipate the different methods students might use for solving the problem.</td>
<td>Plan how you might record students' methods (practice how you will translate student oral strategies into mathematical notation).</td>
</tr>
<tr>
<td>List genuine questions you will ask to fully understand and represent your students' methods, as well as questions to press/scaffold student learning.</td>
<td>Anticipate what you will do if very few strategies are offered and if there are wrong answers and so on. How will you guide your students' thinking?</td>
</tr>
</table>

After the Math Talk Reflection

This is the time for teachers to step back and consider whether the goals of the Math Talk were achieved and think about the level of student discourse that occurred. Use copies of Figure 4.5 found on the Digital Resources (aftermt.pdf). Has it gone well? Are there areas for improvement? Reflect by placing a check mark in the box next to the big idea done effectively. Fill in the blanks with student names or responses to record the context of the lesson. Or, ask a peer or coach to observe and use the form to provide feedback by highlighting observed teacher moves. It is not intended for every box to be checked in every Math Talk. As teachers implement Math Talks, this reflective tool may be used as needed. Eventually, these questions and teacher moves will become second nature.

Note: Students should also reflect after a Math Talk. If students used Figure 4.3 independently, have them write what they learned on the back of Figure 4.3 for an After Math Talk Reflection. Or, for deeper understanding, facilitate a whole group discussion and record student talk on a KWL chart. Students can share what they discovered and their wonderings to self-reflect on their discussions and new learning.

Figure 4.5 After the Math Talk Reflection

Through carefully planned, discourse-rich Math Talks, interactions between teachers and students should feel conversational. While students explain their thinking, the teacher should be genuinely interested in student ideas and assist the class in clarifying and communicating that thinking. Teachers should prompt students to ask questions as the teacher records the process and communicates with the whole class. These reflections will guide the teacher and students to make this time more effective.

Let's take a peek into a middle school classroom where the teacher is guiding students to make connections between their prior knowledge and the new content of solving multi-digit multiplication problems in a Math Talk.

As students settle in, Ms. Kelly writes "24 + 26" on a piece of chart paper.

Ms. Kelly: Put your thumb up when you have a strategy in mind that will help you solve this problem.

She waits and when most students have thumbs up, she calls on Emma.

Emma: I used place value, or breaking apart, and I wrote 20 + 20 is 40 and 6 + 4 is 10.

Ms. Kelly: Did anyone else solve it using Emma's strategy, the breaking apart or place value strategy?

Several students raise their hands to show a connection to Emma's thinking.

Ms. Kelly: I see not all of us solved the problem as Emma did. Who has a different strategy?

She waits until several students have their thumbs up and then calls on Mark.

Mark: I just knew that 6 and 4 made 10, and then I added it to the 20 and got 30 and then added the other 20.

Ms. Kelly: How are Mark and Emma's strategies different? How are they alike? Turn and tell your friend what you are thinking.

As the conversation subsides, she calls on Chris to share what her partner was thinking.

Chris: Well Jolene said both Emma and Mark used the break apart or place value strategy, but Emma started in the tens place, and I started in the ones place.

Ms. Kelly: Can anyone add onto that idea?

She calls on Beth.

Beth: I think Mark was using a known fact of making tens, and he did it quickly in his head to try to be efficient.

Ray: We agree! My partner said both strategies were efficient and made sense. We both agreed we could see what both Emma and Mark did.

Ms. Kelly: Seeing similarities and differences in strategies can help us think about ways to begin to solve a problem. Did anyone else do it another way?

She calls on Dan.

Dan: I took 1 away from the 26 and gave it to the 24. Then, I just knew the double of 25 + 25 was 50.

Ms. Kelly: Can you do that? Will that always work? Turn and talk to your partner about Dan's strategy.

Students talk with partners and share ideas with the whole group.

Ms. Kelly: I have a new problem to challenge you!

She writes 25 × 8 on the chart.

Ms. Kelly: Put your thumbs up when you have a strategy you can share of how you decided to solve this problem.

She waits and then calls on Beth.

Beth: I knew that 4 quarters equals 1 dollar, so I doubled that and got 2 dollars or 200.

Ms. Kelly: Hmm, what did Beth do? How does that work? Does that make sense? Turn and talk to your neighbor. Partner buzz.

Ray: Ms. Kelly, I understand what Beth did but I respectfully chose another way of solving the problem. I did not use money. I used the break apart or partial product way of solving the problem. I said 8 groups of 20 is 160 and 8 groups of 5 is 40 and that was easy for me to see it was 200.

Ms. Kelly: Class, do you agree with Ray's strategy? Can you see how he was thinking? How many of you used Ray's strategy to solve this problem? Did anyone do it another way?

Jace puts his thumb up, and Ms. Kelly asks him to share his strategy.

Jace: Well, my dad showed me to put the big number on top and the small number underneath and then start with the one's place. So I did 8×5, and it gave me 40. And then I did 8×20, which gave me 160. I added $40 + 160$ and got 200.

Ray: Hey Jace, that is the same way I did it, but I started with the big numbers in the tens place! We both used the break apart strategy. It is like expanding the number into two parts.

Ms. Kelly: So, there are many ways to solve our problems that make sense to us! Let's try one more.

She writes $25 \times 26 =$ vertically on the chart and waits for a long time until students begin to put their thumbs up. She calls on Ryan.

Ryan: Well, I multiplied 6×5 and got 30 and 20×20 and got 400. So, 430 is the answer.

Ms. Kelly: Did anyone use that strategy?

A few hands go up.

Ms. Kelly: So, not everyone used Ryan's strategy. Who was thinking differently than Ryan?

She calls on Maria.

Maria: I used my prior knowledge. I knew that 8 × 25 was 200, so I asked myself how many groups of 8 are there in 26. The answer is 3 groups of 8 are in 26, so that gave me 3 groups of 200. 600. But there are still 2 more groups of 25 that I have not used so I added 50 to the 600. I got 650. But that answer is different than Ryan's solution?

Students' thumbs go up. Ms. Kelly calls on Natalia.

Natalia: I knew that 10 groups of 25 were 250. Then, I doubled that to make 20 groups of 25 and that was 500. Then, I thought about 6 groups of 25, and I made a group of four 25s and a group of two 25s. That was easy because I know 4 quarters is 100 and 2 quarters is 50. So, I added 150 to 500 and I got the same answer as Maria.

Ryan: I know what I did. I disagree with myself! I did not think about how many groups of 25! If I think about all the groups of 25, that's a lot of groups…I can see it with quarters now. My solution was too small.

Ms. Kelly: Could we go back to Ryan's strategy by creating a picture or diagram of it to see where the mistake happened or why? Let's puzzle on that until tomorrow's Math Talk time. We have a great opportunity to "grow our brains" right now!

As we listen to the discourse that took place during this Math Talk, we recognize many genuine questions from both the teacher and students. Students justified their thinking as they moved toward the solution. They reasoned about each other's ideas and how they began to clarify their own understanding about the various ways to solve the problem. This models the beauty of discourse in a powerful way. Ms. Kelly realized there was a great opportunity for learning in this Math Talk, and it was her responsibility to begin the next Math Talk with this idea to cement her students' learning, using other students' strategies and exploration of the accuracy of the American algorithm. This

Math Talk took on a life of it's own, and there was no road map to follow. Ms. Kelly needed to think on her feet about questions to ask to continually grow her students as mathematicians.

As demonstrated in Ms. Kelly's classroom, there are several aspects that must be in place to make sure all students get the most from a Math Talk experience. It is critical to make time to:

- Create a safe, risk-taking environment where every student's thinking is valued.
- Build a mathematical community of learners through agreements.
- Select problems or strings of problems that allow access for all students in a learning community.
- Select problems that intentionally highlight mathematical concepts.
- Select problems of various levels of difficulty that can be solved in a variety of ways, using multiple strategies.
- Use concrete models, sketches, and diagrams.
- Give students time to think first and then share.
- Teach students how to check for understanding and set goals.
- Model how to interact student-to-student.
- Learn to disagree with yourself or self-correct ideas.

During Math Talks, students become the creators as well as the beneficiaries of mathematical ideas. Eleanor Duckworth writes in her book, *"The Having of Wonderful Ideas" and Other Essays on Teaching and Learning*, about nurturing student's ideas:

> The wonderful ideas that I refer to need not necessarily look wonderful to the outside world. I see no difference between wonderful ideas that many other people have already had, and the wonderful ideas that nobody has yet happened upon...The more we help children to have their wonderful ideas and to feel good about themselves for having them, the more likely it is that they will someday happen upon wonderful ideas that no one else has happened upon before.
>
> —Duckworth 1987, 14

Conclusion

Engaging in Math Talks can change the classroom culture. Steve Leinwand says it eloquently in his foreword for *Number Talks: Fractions, Decimals, and Percentages*, "Done well, there is something so engaging, so mathematically rich, and yes, even magical about Math Talks" (2016, xxv). Both the simplicity and power of Math Talks, when done well, can help shift teachers' mindsets from *telling* students what to do to *listening* to their thinking and learning with them. When done often, Math Talks fulfill the vision of facilitating mathematical discourse in our classrooms to attain the goal of deeply making sense of the mathematics.

As you engage in Math Talks with your students, there will be many opportunities to marvel at their wonderful ideas.

Reflect and Discuss

1. What is a Math Talk?

2. Think about your daily schedule or math block. How could you incorporate a Math Talk routine into your day?

3. What teacher talk moves do you want to investigate as you orchestrate a Math Talk community of learners?

4. Explore some of the free web resources on page 103 for fostering Math Talks in your classroom.

Chapter 5

Equity and Engagement

Principles to Actions: Ensuring Mathematical Success for All (NCTM 2014) identifies access and equity as essential elements for successful mathematics programs. This principle challenges us as teachers to step back and reflect on how we teach mathematics to ensure students "have access to a high-quality mathematics curriculum, effective teaching and learning, high expectations, and the support and resources needed to maximize every student's learning potential" (NCTM 2014, 59). When thinking about equity versus equality, these pictures in Figure 5.1 should come to mind:

Figure 5.1 Equality and Equity

EQUALITY   EQUITY

Printed with permission from Froehle 2012

Equity does not mean every student should receive identical instruction, tools, or problems; instead, it "demands that reasonable and appropriate accommodations be made as needed to promote access and attainment for all students" (NCTM 2000, 12). Access and equity in the classroom rests on teacher beliefs and the practices used to achieve outcomes in mathematics not connected or predicted by student characteristics. A teacher's vision of access and equity requires the teacher to be responsive to students' backgrounds, experiences, and knowledge when designing, implementing, and assessing the effectiveness of mathematical discourse opportunities in classrooms.

Equitable participation is the hallmark of a Socratic society and democratic discussion, but it does not happen automatically. Teachers need to be willing to let go of the expert/ evaluator role to give students opportunities to take risks and share what they think at that moment. This can be difficult for many teachers. While teachers step back and listen to student ideas and thinking, they must also be proactively monitoring student participation to move toward the goal of the lesson, as well as to ensure all voices are being heard. This challenge can be daunting when teachers begin to facilitate and guide student discussions and embolden students into a meaningful discourse community (Carpenter et. al 2015).

Letting go of control of the lesson can be difficult at first. The teacher may feel the volume in the room has increased. But, the louder environment one might experience may be the result of increased engagement in the learning! It is the teacher's role to be present as discussions happen to make sure this louder learning environment is productively moving toward the mathematical goal of the lesson.

The unpredictability of not knowing how students will respond can be overwhelming at first as you plan discourse opportunities. Learning to listen carefully to student ideas, assess each student's understanding based on their thoughts, and then respond accordingly is truly the art of facilitating discourse in your classroom. At the forefront of this success is choosing and planning tasks that require students to productively struggle.

Mathematics has a beauty all students can experience. However, it is often taught as a performance subject rather than a thinking or learning process. When students engage in discourse, it enables them to see mathematical ideas, make connections, and understand concepts. This beauty can be perplexing, yet inspiring. One of the biggest roadblocks to this goal is the lack of self-confidence that far too many students develop early in their educational paths. This notion leads them to view mathematics as something that is far beyond their grasp, and they begin to see mathematics as being within the reach of only a few exceptional mathematical whizzes. Our adult community may also contribute to this idea by unknowingly reinforcing the notion when we excuse poor student performance as a bad gene that was passed down. As well, educators and schools may reinforce the misconception by sorting students by ability (often called tracking), conveying the belief that only some students can do well in mathematics while others cannot. This can hinder a teacher's ability to nurture a discourse-rich environment at any level.

In her book, *What's Math Got to Do With It? Helping Children Learn to Love Their Least Favorite Subject and Why It's Important for America,* (2008) Jo Boaler writes about students being "stuck in the slow lane" and how American grouping systems perpetuate low achievement in mathematics. Researchers consistently find the most important factor in math success to be what they call "opportunity to learn" (Burris, Heubert, and Levin 2006). When students are not given opportunities to learn from challenging and high-level work, they do not achieve at high levels. In a mixed-ability group, the teacher has to open the math, making it suitable for students working at different levels and different speeds. When teachers assign students to ability-leveled groups, they make decisions that affect their long-term achievement and self-efficacy as mathematicians. One of the most important goals of our educational system is to provide stimulating environments for students. They are environments in which students' interests and curiosity are piqued and nurtured, with teachers who are ready to recognize, cultivate, and develop the potential every student shows at different times and areas. Only in this setting will schools become equitable learning environments in which all students are given the chance to be successful mathematicians.

Effective mathematics instruction addresses students' culture, conditions, and language to enhance mathematical learning on a daily basis for all students. When challenging tasks, discourse, genuine questions, and problem solving are used, the potential to provide deeper learning opportunities for all students occurs. Teachers who demand higher-order thinking and mathematical achievement for all, no matter their background, culture, socioeconomic level, or language proficiency, share the expectation that every student can be a mathematician. Mathematics can be seen as a way to see the world and create empowered students who are ready to think, make sense of, and explore the world they live in. For many students, this is the missing link…how school mathematics is connected to the real world.

Some powerful beliefs and tools that were discussed in previous chapters can be implemented in teachers' daily routines to make mathematics more inclusive for all.

1. **Have high expectations for all students through the use of high-demand tasks.**

A clear way to improve achievement and promote equity happens when more students are offered high-level learning opportunities. Teachers who believe mathematics ability is a function of opportunity, experience, and effort, not of innate intelligence, will share those high expectations and inspire students to fight for sensemaking.

2. **Promote a growth mindset so all students believe they can achieve in mathematics.**

Through mathematical discussions and discourse, students see how working hard, making mistakes, and learning from each other is what mathematicians do. Students will learn to believe they can achieve!

3. **Provide differentiated levels of support within whole-group discussions.**

When students learn in a safe learning environment, they take risks, clarify misconceptions, and celebrate successes with the class. The practice of isolating low-achieving students into slower-paced math groups or high-achieving students into accelerated math groups should be eliminated by offering differentiated levels of support (described in Increasing Engagement with Problem Solving on pages 120–126).

4. **Use strategies to support language acquisition for all students.**

Students who are not fluent in English can learn the language of mathematics at grade level and beyond. Discourse-rich classrooms offer more opportunities to use language.

5. **Teach students how to work and collaborate as a team of learners.**

When students work collaboratively, mathematical tasks give them opportunities to see and understand mathematical connections and differing perspectives. As a result, collaborative tasks offer equitable outcomes. Learning mathematics is a social process, and the classroom setting should reflect this.

Increasing Engagement with Problem Solving

Studies have shown the teacher has a greater impact on student learning than any other variable (Darling-Hammond 2000). The safe environment and inclusive community a teacher builds makes a difference, but there are other critical variables that have a big part in the math classroom learning experience toward problem solving and critical thinking. Those variables are the curriculum, math resources, and tasks through which students learn mathematics. Great mathematical tasks can make the difference between happy, engaged, curious, and inspired learners and disengaged, unmotivated, and indifferent learners. The math tasks, problems, and activities teachers plan and implement on a daily basis should pique students' interest and curiosity. The ideas and tools we talk about throughout this book can be used to facilitate rich mathematical discourse in a classroom and engage students in a way that promotes a continued curiosity and desire to deeply understand mathematical ideas and processes. Different tasks will create a variety of talk, as students puzzle, try a strategy, talk about their thinking, and make connections with a variety of strategies and ideas.

In the majority of classrooms in the United States, students use textbooks which communicate the idea that problem solving is a process for selecting which computational procedure to use. This information is often tacked onto the end of chapters. For instance, "Ellie needs 5 and $\frac{7}{8}$ yards of fabric at $8.59 a yard. What will it cost?" This use of what I call "routine problems" asks students to translate the description of the situation into an equation. The purpose of routine problems is to find the correct solution.

In the 1950s, George Polya (1948) broadened our sense of problem solving by describing strategies used by his college students. He stated, "A problem is not a problem if it can be solved in less than 24 hours." His discussion presented the idea that good problems shouldn't have readily accessible procedures, which lead students in one specific direction to a specific solution.

Consider a student solving this type of problem found in many textbooks, "Dog treats cost $3.98 per bag. How much would you pay for 12 bags?" If the student solving this problem understands there are 12 bags and each costs the same ($3.98), then he or she will see the answer can be found by repeated addition or multiplication. If this is the case, the teacher should be asking "Is this a 'problem' for my students?" In this case, the answer is no. It is a drill exercise, which can be valuable, but it is not a problem. A problem is a question in which students critically think and reason.

Some textbook companies provide a few problems that require students to do more than merely determine which operation to use and provide a set answer. These problems are sometimes called non-routine or process problems in which thinking is required and multiple processes or strategies can be used to find a solution.

Consider the following problem: "How many different ways can you make change for a quarter?" There is no obvious path or computational procedure to use. The student has to work it out. Take a moment to try the problem on your own. How many ways did you find? How did you work it? Did you make a list? Did you draw a picture? Did you make a table? Did you use prior knowledge? Did you work with a friend? If you did any of these, you were using problem-solving strategies. If you found all twelve, how did you organize your data to be sure you found them all? Strategies can help students find solutions and provide a pathway to understand the problem.

By the early 1980s, the idea of strategies found their way into classrooms. Many textbooks exactly taught students a few strategies. But the idea of just teaching strategies to students was not creating the depth of understanding hoped for. So, what was missing? It seemed to be the connection between language, mathematics, and cognitive principles.

In their book *How Students Learn* (2005), Donovan and Bransford share three main principles to synthesize a tremendous amount of information about connecting learning to human cognition. They are:

1. Engaging prior understandings (using prior knowledge, confronting preconceptions and misconceptions)

2. Organizing knowledge (developing a deep foundation of factual knowledge organized into coherent conceptual frameworks that reflect contexts for application and knowing when to use which information—referred to as conditionalized knowledge)

3. Monitoring and reflecting on one's learning (developing metacognitive processes and self-regulatory capabilities)

Thinking about the three principles above, one can see the importance of connecting to prior knowledge, activating relevant contexts, building organized knowledge, and developing self-awareness as a learner. Math needs to be expressive, spoken, written, responsive, interactive, dynamic, creative, playful, clarifying, defining, and scaffolded.

There are two basic types of connections that will help students achieve a community of learners: contextual connections and conceptual connections. Contextual connections occur when teachers ask questions that stimulate students' thinking about the situation or context of a math problem. When students connect new ideas to their prior knowledge and experiences and learn from their preconceptions and misconceptions, they are connecting math to their self-concepts. Often, during this metacognition, students might ask themselves: *What does this situation remind me of? Have I ever been in any situation like this? What do I already know about this situation? How will what I know help me begin?* And so on.

Students' connections are not limited to their homes. Conceptual connections occur when students connect new mathematical concepts to other concepts or big ideas within and across the various strands of mathematics. They can also connect new ideas to related procedures, visual models, and representations. Some of the questions students might ask themselves are: *What is the main mathematical idea here? Where have I seen this mathematical idea before? How is this idea related to another concept we already studied? Can I use my prior knowledge to help me with this problem?*

According to *Principles to Actions: Ensuring Mathematical Success for All:*

> **Effective teaching of mathematics engages students in solving and discussing tasks that promote mathematical reasoning and problem solving and allow multiple entry points and varied solutions strategies.**
> **—NCTM 2014, 17**

Tasks that promote reasoning and problem solving require students to engage in mathematical thinking and sensemaking. Therefore, the first step toward facilitating high-quality mathematical discourse in the K–12 classroom involves choosing tasks that are worth talking about. Consider these two simple tasks:

1. To which fact family does the fact $4 \times 5 = 20$ belong?

2. Describe the picture/diagram with an equation.

$$\bigstar \ \bigstar \ \bigstar \ \bigstar \ \bigstar$$
$$\bigstar \ \bigstar \ \bigstar \ \bigstar \ \bigstar$$
$$\bigstar \ \bigstar \ \bigstar \ \bigstar \ \bigstar$$
$$\bigstar \ \bigstar \ \bigstar \ \bigstar \ \bigstar$$

The first task has students demonstrate one type of thinking but does not ask them to make connections or engage in mathematical discourse. If students do not know or understand the definition of the term *fact family*, they cannot engage in the problem or show what they know. The second task provides multiple entry points and allows students to show what they know in ways that will spark meaningful mathematical discourse. The language within the question is not dependent on specialized mathematical language, which addresses equity, offering all students the opportunity to show what they know.

One of the most important things a teacher can do for all learners is ensure they have opportunities to enjoy the puzzling aspect of mathematics at a level at which students can be successful. The second task demands engagement with concepts, offering the possibility for students to make connections and prompting them to share strategies with fellow classmates. Providing rote, one-answer tasks never takes students on the path to truly enjoying that thought-provoking puzzle of math—even if they do enjoy the reward of being correct.

When teachers consider whether a task is worth talking about, they might ask the following two questions:

1. **Is the task a high-cognitive demand task that provides multiple access/entry points?**

   The cognitive demand of a task is the level of cognitive engagement needed to complete the task (Stein et al. 2009). Problems that require only memorization are at the low end of cognitive demand, whereas tasks that require students to make connections between and among mathematical ideas in new ways are considered high-cognitive demand tasks. Low-demand tasks do not promote student-to-student discourse and often lead students to believe it is their jobs to give answers in mathematics and, hopefully, those the teacher wants. Low-demand tasks do not stimulate enough mental engagement to produce a rich mathematical discourse community for all learners.

2. **Does the task support the mathematical goal of the lesson by making connections and promoting a depth of knowledge?**

   Choosing a task that matches the mathematical goal of a lesson is another high-leverage practice and is closely related to facilitating a purposeful mathematical discussion. It is important for teachers to keep this goal in mind when various solution strategies are shared because, when students begin to share their ideas, it can increase the odds of the main point of the lesson being lost.

Characteristics of Mathematical Instructional Tasks (Smith and Stein 1998, 348) in Figure 5.1 on pages 125–126 offers clear guidelines for examining the cognitive demands of mathematical tasks that continue to benefit teachers as they choose tasks to support the goals of their lessons and promote meaningful discourse opportunities in their classrooms.

Figure 5.1 Characteristics of Mathematical Tasks

Levels of Demand

Lower-level demands (memorization):

- Involve either reproducing previously learned facts, rules, formulas, or definitions or committing facts, rules, formulas, or definitions to memory.

- Cannot be solved using procedures because a procedure does not exist, or because the time frame in which the task is being completed is too short to use a procedure.

- Are not ambiguous. Such tasks involve the exact reproduction of previously seen material, and what is to be reproduced is clearly and directly stated.

- Have no connection to the concepts or meanings that underlie the facts, rules, formulas, or definitions being learned or reproduced.

Lower-level demands (procedures without connections):

- Are algorithmic. Use of the procedure either is specifically called for or is evident from prior instruction, experience, or placement of the task.

- Require limited cognitive demand for successful completion. Little ambiguity exists about what needs to be done and about how to do it.

- Have no connection to the concepts or meanings that underlie the procedure being used.

- Are focused on producing correct answers instead of developing mathematical understanding.

- Require no explanations or explanations that focus solely on describing the procedure that was used.

Higher-level demands (doing mathematics):

- Require complex and non-algorithmic thinking—a predictable, well-rehearsed approach or pathway is not explicitly suggested by the task, task instructions, or a worked-out example.

- Require students to explore and understand the nature of mathematical concepts, processes, or relationships.

- Demand self-monitoring or self-regulation of one's own cognitive processes.

- Require students to access relevant knowledge and experiences and make appropriate use of them in working through the task.

- Require students to analyze the task and actively examine task constraints that may limit possible solution strategies and solutions.

Higher-level demands (doing mathematics): (cont.)

- Require considerable cognitive effort and may involve some level of anxiety for the student because of the unpredictable nature of the solution process required.

These characteristics are derived from the work of Doyle on academic tasks (1988) and Resnick on high-level thinking skills (1987), the Professional Standards for Teaching Mathematics (NCTM 1991), and the examination and categorization of hundreds of tasks used in QUASAR classrooms (Stein, Grover, and Henningsen 1996; Stien, Lane, and Silver 1996).

Printed with permission from Smith and Stein 1998

If teachers reference these levels, they will be more aware of what the cognitive demand tasks are requiring and see whether they match the learning goals for the chosen lesson. When math tasks are designed so students can find different ways to explore strategies and methods, students can make mathematical connections. Then, mathematics can become equitable and full of sensemaking opportunities for all.

In her book, *Mathematical Mindsets: Unleashing Students' Potential through Creative Math, Inspiring Message and Innovative Teaching,* Jo Boaler (2016, 90) discusses the *how* of changing tasks from being closed (i.e., fixed mindset tasks) to growth mindset tasks that offer students the space to struggle, make mistakes, and learn. She offers five suggestions, which can work to open math tasks and increase the potential for depth in mathematical learning:

1. Open up the task so there are multiple methods, pathways, and representations;
2. Include inquiry opportunities;
3. Ask the problem before teaching the method;
4. Add a visual component, and ask students to show they see the mathematics;
5. Extend the task to make a lower floor and higher ceiling; and
6. Ask students to convince and reason. Be skeptical.

Boaler explains that students deserve a mathematically rich environment to discuss, argue, and ask questions while making sense of mathematics. When they see the beauty of mathematics, they will truly be motivated to engage in the learning process and understand what it means to be a mathematician. This all begins with choosing an excellent task.

Creating Engaging Tasks

As teachers analyze mathematical tasks offered in various resources using Smith and Stein's Level of Demands (Figure 5.1), they may find it necessary to alter problems to make them more demanding to solve. Susan O'Connell, in her book *Math in Practice: A Guide for Teachers* (2016), guides teachers to think deeply about creating tasks that tie problem solving to the real world. Marian Small, in her book *Good Questions: Great Ways to Differentiate Mathematics Instruction* (2012), provides two powerful and universal strategies teachers can use with any math content: Open Questions and Parallel Tasks. Here are a few examples of how contextualizing the problem and creating open-ended questions can be used with any resource.

Personalize the Context

Students are motivated to engage and think when the context of a problem appeals to them. Initially, they are much more interested in the context than the mathematics within the problem. This is a good thing. Students need to learn to step back and think deeply about the context of story problems. When students consider many perspectives from one particular problem, they create meaning and can begin searching for patterns and relationships, which is what mathematics is, at its core. So, the good news is we can use the innate, pattern-creating, sensemaking, and inductive reasoning students bring to school with rich problem-solving tasks to connect contextual and conceptual understanding. Beginning with an engaging task is the start!

Many students do not see mathematics in the world around them. They see it as a subject in school, which has no purpose outside the four walls of the class. But, teachers have to hook students into wanting to do the math. It can be as simple as adding student names from the classroom, your principal or librarian's name, a local restaurant's name, the name of the local skate park, or the local baseball team into tasks to engage students in learning.

Create Open-Ended Questions

One of the most important things a teacher can do is ensure students have opportunities to enjoy the puzzle aspect of mathematics and at a level at which they can succeed. Providing only rote memorization tasks will never give students opportunities to truly enjoy mathematics, even if they do enjoy the success of being correct. When creating open-ended questions, teachers must know the big idea of the mathematics lesson and problems that allow students choice. These are some ways to create open-ended questions:

- Be mathematically meaningful to students;
- Include single questions or tasks that are inclusive; and
- Include parallel tasks, which are accessible to all students.

Here are two examples of close-ended questions:

1. There are 583 students in Amy's school in the morning. Then, 99 of the fifth graders leave for a field trip. How many students are left in the school?

2. There are 61 third grade students in Amy's school. Then, 19 of them go to the library to work on their research projects. How many are left in the classroom?

Here are two examples of open-ended questions, which have applied the characteristics of open-ended questions from page 127:

Option 1: Are there more multiples of 3 or more multiples of 4 between 1 and 100? How many more? How did you know?

Option 2: Are there more prime numbers or more composite numbers between 1 and 100? How many more? How did you prove it?

How was this done? There are several strategies to change questions from closed to open-ended. It is important to see these changes as a change in approach rather than a change in question content.

Turning around a Question

Closed: What is half of 20?

Open: The number 10 is a fraction of a number. What could the fraction be?

Asking about Similarities and Differences

Closed: How is the number 7 similar to 99? How is it different?

Open: What do you notice about the numbers 7 and 99? How do you know? Can you convince us?

Removing Some Numbers

Closed: If there are 25 students in Ms. Kelly's classroom and 31 in Ms. Lauren's classroom, how many are there altogether?

Open: There are 56 students in fifth grade at Mariposa School. How many could be in Ms. Sampson's room and Mr. Hill's class?

Changing the Question

Closed: Rodney has 4 packages of pencils. There are 6 pencils in each package. How many pencils does Rodney have?

Open: Rodney has some packages of pencils. There are 2 more pencils in each package than the number of packages. How many pencils might Rodney have?

Reaching Out with Resources

There are also so many excellent websites and search engines available that can share these types of excellent tasks to K–12 teachers. Here are a few to explore when looking for high-demand tasks to use with your students. Search for the activities on the websites listed below to find the open-ended questions that work for your students.

- **Graham Fletcher's 3-Acts Lessons** gfletchy.com/3-act-.lessons/
- **Estimation 180** www.estimation180.com
- **Number Strings** numberstrings.com
- **Mathalicious; grades 6–12** www.mathalicious.com
- **Mathalicious Assessment Resource Service (MARS)** map.mathshell.org/materials/index.php
- **dy/dan from Dan Meyer** blog.mrmeyer.com
- **NCTM Illuminations** illuminations.nctm.org
- **Divisible by 3 from Andrew Stadel** mr-stadel.blogspot.com
- **Which One Doesn't Belong (WODB)** wodb.ca
- **Robert Kaplinsky** robertkaplinsky.com/lessons
- **Visual Patterns: grades K–12** www.visualpatterns.org
- **NCTM The Math Forum** www.mathforum.org
- **Problems of the Month—inside mathematics** www.Insidemathematics.org/problems-of-the-month/download-problems-of-the-month
- **Illustrative Mathematics** www.IllustrativeMathematics.org
- **The Classroom Chef** www.classroomchef.com/chefs
- **Jo Boaler, Youcubed** www.youcubed.org
- **NRICH** nrich.maths.org

- **Yummy Math** www.yummymath.com
- **desmos™** teacher.desmos.com
- **Math Munch** mathmunch.org
- **Transformulas from Jed Butler** transformulas.org
- **Gifsmos from Chris Lusto** www.gifsmos.com
- **Database of Math Activities in Desmos** bit.ly/desmosbank
- **Database of Activity-Based Learning** bit.ly/MTBoSbank
- **Robert Kaplinsky's Problem-Based Search Engine** bit.ly/RKsearchengine
- **TMathC** tmathc.com
- **Geogebra Creations from Dr. Ted Coe** tedcoe.com/math
- **The MTBoS Search Engine** bit.ly/MTBOSS

Eliciting Student Thinking

Once an excellent task is chosen, teachers can focus on eliciting student thinking. Teachers do this by consistently asking students to share thinking and planning strategies. A teacher must be prepared to ask genuine questions that encourage students to communicate their ideas during the thinking process. Again, genuine questions are questions that do not elicit an expectation of a correct or incorrect answer from the student. What teachers are looking for when creating genuine questions is to frame the question to include partial thinking. When teachers ask questions about correct, incorrect, and partial thinking or incomplete strategies, they model how to listen, observe, and reason. It is important for teachers to not impose or share their own ideas, strategies, or solutions with their students while asking questions. Without genuine questions, teachers will see students anticipate the teacher's expectations. They think their ideas were not the best. The classroom becomes a place that promotes the idea that doing math is to get to the teacher's ideas or solutions instead of the student's. This will hinder the level of discourse in the classroom and reallocate the learning out of students' hands—something we as educators don't want to do.

But once students readily share their ideas with classmates and the teacher, another challenge begins. Students must interact with each other's ideas. This involves more than listening. It requires students to notice, understand, and evaluate each other's strategies. Teachers can plan to support this important next step. In their book *Children's*

Mathematics: Cognitively Guided Instruction (Carpenter et. al 2015), the authors share three levels of categorizing students' engagement in other students' ideas:

1. Comparing an idea to another student's idea.
2. Attending to the details of another student's idea.
3. Building on another student's idea.

Their research shows that students who participate in all three levels of engagement have the highest levels of mathematical achievement. Teachers must anticipate students' knowledge of the different concepts and plan genuine questions for every lesson to support these three levels of engagement.

Conclusion

It is the teacher who chooses and creates high-demand tasks that students will use to develop problem-solving abilities. Teachers set high expectations, orchestrate feedback to support students' shifts in mindset, provide strategies for students to access the language of mathematics, and support social and collaborative processes when learning together. To provide access and equity, teachers must move beyond "good teaching" to facilitate learning experiences that offer students of all learning levels and strengths opportunities to engage successfully in the classroom and to collaborate to learn challenging mathematics. Providing opportunities to express, discuss, question, and argue is important when teaching students to fight for sensemaking in mathematics.

Reflect and Discuss

1. How can you determine whether a math task is worth talking about?

2. Find a routine task in your textbook or resource and think about how you could modify it to be a non-routine task.

3. Choose a problem-solving task you have used and use the Levels of Demand on pages 125–126 to determine which level of demand it meets. How can you alter the task to have a higher or lower cognitive demand to provide access for all students?

4. What are some discourse strategies you can incorporate into your instruction to provide access and equity for all students?

Chapter 6

Getting Kids Ready to Talk! The First 20 Days of Discourse

Teachers spend a lot of time setting up their physical classrooms to nurture student learning. It is important to take the time to consider the math environment you wish to create and share with your students as they enter the room. Do you have genuine math questions around the room? Have you created a bulletin board or special space for math vocabulary, learning outcomes, samples of student work? Are your manipulatives organized and stored in a place for easy access for students at anytime during your math block? How are your desks organized to promote student-to-student discourse? Have you organized a space to store math journals so you and your students can easily access them? Do you have space to put up posters of practice and process standards or mathematical strategy posters like the ones found in Figures 6.1 and 6.2 on page 134? Will your students know the high expectations you have for them as mathematicians as they walk through the classroom door? All of these questions will set you in a positive and proactive direction that will engage all students to talk about math on Day 1!

The First 20 Days of Discourse is a guideline to support the importance of building community and establishing norms to promote rich discourse in K–12 classrooms. Treat each day as a suggestion that can be used in conjunction with normally scheduled content. The daily activities are aligned to support mathematical practice and process standards found in state and national standards. CCSS's Standards for Mathematical Practice and the Texas Essential Knowledge and Skills for Mathematics (TEKS-M) Mathematical Process Standards are shown as examples.

Figure 6.1 Example of Mathematical Practice Poster, K–2

Printed with permission from the Math Learning Center 2015a

Figure 6.2 Multiplication Fact Strategy Poster

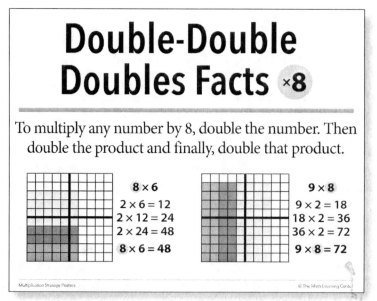

Printed with permission from the Math Learning Center 2015b

Day 1: Building a Mathematical Community

Preparing for mathematical discourse can be an easy and flexible process that can be woven into already established schedules and lesson plans. Begin simply by developing classroom agreements (norms), such as the What Do Mathematicians Do? poster below, during Day 1 of mathematics instruction. Discuss the importance of working and learning together as a team of mathematicians and introduce the idea of how anchors charts, such as community agreements, student work criteria, and instructional posters, can assist students in gradually becoming mathematicians. Share with students some things mathematicians do.

What Do Mathematicians Do?

Look for more than one way to solve a problem.

Organize thinking and strategies.

Look for other ideas (patterns and relationships) in the mathematics.

Meet and share ideas with other mathematicians.

Give feedback to others to help them improve their thinking.

Look for new ideas in other people's thinking.

Puzzle about new ideas and work together to make connections.

Ask each other to prove their thinking. (How do you know?)

Prove ideas or strategies work by using another strategy
to get to the same result or conclusion.

Listen to different ideas and think about making sense of them.

Ask genuine questions about different math ideas.

Persevere and try new things.

Take risks, make mistakes, and learn from your experiences.

Support each other.

Fight for sensemaking!

Begin to gather and create a list of agreements, like the ones on page 135. You might title your agreement "Being a Mathematician: Our Community Agreements" or "Our Math Community Agreements." Make sure statements in the agreements are phrased in the positive, using student language that shows what to do rather than what not to do. Anchor charts should be easily accessed throughout the year so the class can reference the charts when needed. In conjunction with forming community agreements, this is a great time to introduce mathematical practice and process standards MP1 and MP3 or TEKS-M 1.E.

Sample Activities/Ideas:

After beginning a chart of community agreements, you might use one more of the following activities to celebrate risk taking and discourse in the math classroom.

MP1 Make sense of problems and persevere in solving them. MP3 Construct viable arguments and critique the reasoning of others.	TEKS-M 1.E Create and use representations to organize, record, and communicate mathematical ideas.
Elementary Ideas	
• Which One Doesn't Belong? (Shape or Numbers) wodb.ca/numbers.html • **Quick Write:** What does it mean to do mathematics? • **Exit Ticket:** As a mathematician, I want to…	
Secondary Ideas	
• Which One Doesn't Belong? (Graphs and Equations) wodb.ca/graphs.html • Have students create a journal cover with magazine pictures of people using mathematics: What do mathematicians do? • **Quick Write:** What does it mean to do mathematics? • **Exit Ticket:** What I remember about learning math in elementary school is…	

Day 2: Exploring and Managing Mathematical Tools

Mathematicians use math tools to think deeply, plan strategies, and solve problems. Students need to become familiar with math tools in the classroom and discover how they can promote math talk and reasoning using various strategies. Discuss how mathematical materials can be used as tools to explore, discuss, and problem solve.

Introduce manipulatives, tools, and technology that will be used throughout the year. Where are they stored? How and when are these accessed? How are they put away? Co-construct a list of agreements about appropriate use of tools. Add these ideas to the class agreement list or chart. Connect today's discoveries with mathematical practice and process standards MP3 and MP5 or TEKS-M 1.C.

MP5 Use appropriate tools strategically.	TEKS-M 1.C Select tools, including real objects, manipulatives, paper and pencil, and technology as appropriate, and techniques, including mental math, estimation, and number sense as appropriate, to solve problems.
MP3 Construct viable arguments and critique the reasoning of others.	

Elementary Ideas

- Introduce and have students explore a few new manipulatives or math tools for your grade level content, such as polydrons, ten frames, number racks, ten strips, hundreds grids, base-ten blocks, number lines, and digital manipulatives.
- Collect data and graph information about your class (e.g., lunch count, favorite sport, love/like/so-so about math).

Secondary Ideas

- Create a get-to-know-us glyph or graph to promote discourse. Have students provide similarities and differences that show learning styles and write them on the glyph or graph.
- Introduce and have students free-explore a few new manipulatives, or math tools for your grade-level content, such as tablets/laptops, base-ten blocks, fraction blocks, math apps, compasses, protractors, graphing calculators, and graphing apps (e.g., desmos).
- Journal prompt: How can math tools help us solve tricky problems?

Day 3: Engaging in Accountable Talk

On Day 3, ask students to discuss what good speaking looks and sounds like. The lists on pages 138–139 provide characteristics for these routines during partner or group work in mathematics. This is called accountable talk. Students need to understand they share responsibility for learning, and their talk can spark others to share information as well as to learn new strategies. Tie routines to mathematical practice and process standards MP3 and MP6 or TEKS-M 1.A.

Peer-to-Peer

- Make eye contact with your peer.
- Make gestures, such as nodding or facial expressions, that show you understand their idea.
- Take notes about the ideas you hear.
- Ask thoughtful and genuine questions when your peer is finished talking. Use genuine question posters, if needed.

Small Group

- Make eye contact with the person speaking in your group.
- Use gestures, such as nods or facial expressions, that show you understand the ideas of the person speaking.
- Take notes about the ideas you hear.
- Ask challenging or clarifying questions when the speaker is finished.
- Avoid side conversations with others in the group, focus on the speaker, and let others interact with the speaker.
- Use positive body language, such as sitting up or facing the speaker.
- Use facial expressions to show understanding and active listening.

Whole Group

- Make eye contact with the person speaking.

- Use gestures, such as nods or facial expressions, that show you understand the ideas of the person speaking.

- Take notes of the ideas you hear.

- Ask challenging or clarifying questions when the speaker is finished.

- Avoid side conversations with others in the group. Focus on the speaker and let others interact with the speaker.

- Use positive body language, such as sitting up or facing the speaker.

- Use facial expressions to show understanding and active listening.

- Make sure all voices are heard and that you are listening to ideas different from yours.

- Think about how you could add to ideas from your peers.

MP3 Construct viable arguments and critique the reasoning of others. **MP6 Attend to precision.**	**TEKS-M 1.A Apply mathematics to problems arising in everyday life, society, and the workplace.**

Elementary Ideas

- Role-play how to engage in accountable talk and add to your Community Agreement anchor chart.
- Choose an agreement and ask students to gather evidence about their ability to maintain expectations. What did they do? How can they improve?

Secondary Ideas

- Role-play how to engage in accountable talk with partners, in small groups, and during whole group discussions using the individual or group moves above. Add new agreements to Community Agreement posters.
- Have students choose an agreement and record it in their math journals. Ask them to record evidence of how they maintain that expectation this week as they reflect on their learning community and accountable talk moves.

Day 4: Expectations for the Active Listening in Math

On Day 4, build on yesterday's activities and model of accountable talk, focusing on what active listeners look and sound like. Mathematical conversations and discourse look and sound differently in various-sized groups. How will students communicate to their peers when they are listening and making sense of what they are sharing? Use the Whole Group list on page 139 for characteristics of accountable talk. Introduce mathematical practice and process standards MP3 and MP6 or TEKS-M 1.A.

MP3 Construct viable arguments and critique the reasoning of others. **MP6 Attend to precision.**	**TEKS-M 1.A Apply mathematics to problems arising in everyday life, society, and the workplace.**
Elementary Ideas • Role-play the active listening actions with students and discuss how it feels to be listened to. You might add an agreement to your community agreement anchor chart about active listening and accountable talk. • Ask students to choose new agreements to work toward together. Create agreements as a class.	
Secondary Ideas • Role-play the active listening actions listed above with students and discuss how it feels to be listened to. You might add an agreement to your community agreement anchor chart about active listening and accountable talk. • As a class, choose one agreement to be the focus for the week. Gather evidence and recognize students work toward the agreement.	

Day 5: Using Mathematically Precise Language

On Day 5, have students think about what they precisely mean and how they use mathematical vocabulary accurately. Specialized vocabulary is used in many mathematics lessons. Collect word banks and make vocabulary explicit and embedded in instruction. Many mathematical terms have their meanings in everyday language. For example, we go to baseball diamonds not baseball rhombuses. We play tic-tac-toe and win with three in a row. But really we have rows, columns, or diagonal lines.

Students should explain their thoughts in complete sentences. This helps students develop reasoning and justify their mathematical ideas. Journals and models, such as Cornell Notes and Frayer Models, support vocabulary development. Introduce mathematical practice and process standards MP3 and MP6 or TEKS-M 1.G.

MP3 Construct viable arguments and critique the reasoning of others. **MP6 Attend to precision.**	**TEKS-M 1.G Display, explain, and justify mathematical ideas and arguments using precise mathematical language in written or oral communication.**

Elementary Ideas

- Provide students with the lists *Tricky Words in Mathematics* (page 44) and *Strategies That Support Learning Math While Learning English* (pages 47–48) with graphic organizers to connect math concepts to multiple meaning words.
- Have students use the Free App "Math Vocabulary Cards" www.mathlearningcenter.org/apps.

Secondary Ideas

- Use the Frayer Model or Cornell Notes to have students record mathematical words, pictures, and definitions.
- Create a word resource bank in a student journal. Include important vocabulary for the grade level. Have students write and draw their own pictures and definitions. Review journals often and add to them as students' understanding deepens.

Day 6: Writing/Representation in Math Class

Mathematicians write about their thinking to learn, process, self-assess, and communicate their thinking to others. They also use and record mathematical representations to interpret and model everyday life activities and engage in discourse.

One way to encourage students to write is to use a daily journal. In math class, students will be expected to record and journal about their mathematical thinking on a daily basis. The teacher will read journal entries and respond with questions and "Ahas" that they observe on a weekly basis. This makes journals interactive and creates a place where students converse back and forth with the teacher throughout the year. These interactive math journals will be students' mathematical tools to keep chronological records of their ideas, questions, and discourse. Introduce mathematical practice and process standards MP3 and MP6 or TEKS-M 1.A.

MP3 Construct viable arguments and critique the reasoning of others. MP6 Attend to precision.	TEKS-M 1.A Apply mathematics to problems arising in everyday life, society, and the workplace.

Elementary Ideas

- Introduce interactive math journals and explain that they will be an integral part of the math routine this year.
- Play a favorite math game with students, and have them reflect on their math learning in their journals at the conclusion of the game.

Secondary Ideas

- Introduce interactive math journals and explain that they will be an integral part of the math routine this year.
- Share pictures or images of math journals of famous mathematicians. Discuss the interactive nature of a journal and how others learn from each other when they read and reflect upon each other's writing, and find mistakes or new ideas.
- Give students a journal prompt. Have them respond in their journals and share with a peer. This is a good time to use a feeling prompt, such as "How do you feel about math?" or "When did you use math in the real world this past month?"

Day 7: Using Protocols to Facilitate Discourse

On Day 7, begin the lesson by telling students they will learn protocols that will help provide equity during mathematical discussions. Begin with a protocol to encourage conversations about what equitable discussions look and sound like. This will put accountable talk and active listening habits of mind into action.

Explain how to express and respond to differing ideas and maintain this expectation throughout mathematics class. A goal for using protocols is to eventually have students internalize the process and learn to collaborate without teacher support.

Discuss and role-play sentence frames and starters to assist students at all grade levels to become accountable for their talk and make sure that all students are sharing ideas. Introduce mathematical practice and process standards MP1 and MP4 or TEKS-M 1.G. Below are examples of student sentence frames to smoothly incorporate any protocol.

Expressing Your Ideas and Asking for a Response

- I am thinking…
- It seems to me…
- I think I can add on to your idea…
- You might not agree with me, but here is what I am thinking…
- What do you think?
- Were you thinking differently?

Responding to Your Peer's Ideas

- That is really interesting.
- I had not thought about it that way.
- I have a different idea than you.
- Maybe we should try…
- Why don't we…
- I was wondering…
- What if?…

MP1 Make sense of problems and persevere in solving them. MP4 Model with mathematics.	TEKS-M 1.G Display, explain, and justify mathematical ideas and arguments using precise mathematical language in written or oral communication.

Elementary Ideas

- **Stand Up, Pair Up:** Give each student a picture postcard. Challenge them to think about how the picture relates to learning mathematics. Have students pair up with a partner. Have Partner 1 share while Partner 2 asks clarifying questions or add ons. Then, have partners switch roles. Give students a hand signal after a few minutes to repeat the protocol with a new partner. For more ideas and protocols, see *The Power of Protocols: An Educator's Guide to Better Practice, Third Edition.*

- **Think-Pair-Share:** Write a math statement on the board, and have students think about what it means to them. Before students share, ask them to think about it privately. Then, have students turn and share their thoughts with a partner. Ask three or four students to share their ideas or their partners' ideas.

Secondary Ideas

- **Final Word:** Write a statement, such as *Learning mathematics is a social activity* or *It is important to use a variety of methods to achieve computational fluency.* Give students think time to decide whether they agree or disagree. In small groups, have one student share his or her ideas while the rest of the group listens without commenting. Then, have the next student do the same thing. Finally, have the last student share his or her ideas. Now, have the entire group discuss ideas and actively listen. For more ideas and protocols, see *The Power of Protocols: An Educator's Guide to Better Practice, Third Edition.*

- **Think-Pair-Share:** Write a math statement or mathematical equation on the board. Have students think about what the statement means or whether it is true or false. Then, have them turn and share their thoughts with a partner. Ask three or four students to share their ideas or their partners' ideas.

Day 8: Making Connections in Math Class

Mathematicians wonder how mathematics connects and relates to the world around them, and they make connections between previously used and newly introduced concepts. On Day 8, guide students to see relationships between quantities, mathematics concepts, and strategies as well as how mathematical relationships connect to the world around them. These connections are powerful tools to promote rich discourse during mathematics class. Introduce mathematical practice and process standards MP2 and MP4 or TEKS-M 1.D.

MP2 Reason abstractly and quantitatively. **MP4 Model with mathematics.**	**TEKS-M 1.D Communicate mathematical ideas, reasoning, and their implications using multiple representations, including symbols, diagrams, graphs, and language as appropriate.**

Elementary Ideas

- Play a game using a ten frame or number rack to compose and decompose numbers. Help students make connections between adding and subtracting as you record number sentences for the strategies students used.
- Use models (e.g., base-ten blocks) to help students explain division parts. Students should identify that the quotient is a missing factor, the known factor is the divisor, and the dividend is the product of a multiplication problem.
- Have students generate lists and/or models relating to real-world examples to mathematical concepts.

Secondary Ideas

- Discuss with students how operations with fractions are the same as operations with whole numbers. Have them create diagrams to show the relationship.
- Use a hundred grid divided into quadrants to depict how percentages are simply hundredths. Students should be aware that these grids are another way to justify that fractions and decimals are the same quantity.
- Have students generate lists relating real-world examples and mathematical concepts.

Day 9: Strategic Use of Tools

On Day 9, have students explore why it is important to build with manipulatives and know how to choose the best tools. Students use manipulatives to create concrete representations of concepts to make sense of process while working toward a solution. Provide a math tool (e.g., number rack, base-ten blocks, geoboard, or fraction bars) to talk about how building helps show math conceptually. Add to your community agreement anchor chart: *What does it look, sound, feel like when we use tools appropriately?* Discuss how different tools are used for different mathematical ideas. Introduce mathematical practice and process standards MP5 and MP6 or TEKS-M 1.C.

MP5 Use appropriate tools strategically. **MP6 Attend to precision.**	**TEKS-M 1.C Select tools, including real objects, manipulatives, paper and pencil, and technology as appropriate, and techniques, including mental math, estimation, and number sense as appropriate, to solve problems.**

Elementary Ideas

- Have students explore various math tools, such as pattern blocks, number racks, base-ten blocks, tiles, apps, and tablets. Discuss what students discovered while exploring these tools. Then, shift the focus to mathematical concepts and how to choose math tools to solve problems. Talk about how math tools will be used as "tools not toys."

- Use a clothesline number line to order pictures of numbers in five or ten frames. Focus on students sharing why they chose to place the numbers where they did. Ask students to ask clarifying and challenging questions throughout the process. Show equivalency on the clothesline by clipping numbers together with clothespins.

- Use a clothesline number line to order a set of numbers, such as 0, 10, 20, 21, 31, 100, 75, 80, 90, and 83. Challenge students to share why they placed the numbers in certain places. Ask students to ask clarifying and challenging questions.

MP5 Use appropriate tools strategically. MP6 Attend to precision.	TEKS-M 1.C Select tools, including real objects, manipulatives, paper and pencil, and technology as appropriate, and techniques, including mental math, estimation, and number sense as appropriate, to solve problems.

Secondary Ideas

- Have students explore with tools, such as pattern blocks, geoboards, protractors, base-ten blocks, calculators, decimal grids, number lines, apps, and tablets. Have students discuss what they discovered while exploring these tools. Then, shift the focus to the mathematical concepts and how to strategically choose tools to solve problems. Talk about how they will be used as "tools not toys."

- Use a clothesline number line to order a set of numbers. For example, 0, $\frac{1}{2}$, 1, $1\frac{1}{2}$, $\frac{4}{3}$, $\frac{5}{8}$, $\frac{7}{8}$, 3, $2\frac{3}{4}$, $2\frac{1}{2}$ or x, $2x$, $x + 1$, $x - 2$, $x - 3$, $3x$. Focus on students sharing why they chose to place the numbers where they did. Ask students to engage in clarifying and challenging questions with each other throughout the process.

- Explore equivalent fractions with pattern blocks or geoboards and focus on the picture and relationships that these visual models offer.

- Use pattern blocks to explore, make conjectures, and prove angle measurement.

Day 10: Tools to Sketches to Tell the Math Story

On Day 10, focus on the tools that students have available. Ask them to answer, "How can these tools help me learn more about math?" Have students use the visual models to stay organized and use tools, such as a record sheet or journal entry, to easily retrieve information when needed.

Have students begin a problem. Stop them during partial thinking and hold a gallery walk for students to visit each other's desks. Have students observe sketches and diagrams that tell the story. This will help students see why mathematicians sketch or diagram their thinking. Then, reconvene the class and introduce mathematical practice and process standards MP7 and MP8 or TEKS-M 1.D. Discuss that as patterns reoccur and visual models become easier to see, students will see structures make generalizations within the mathematics concepts.

MP7 Look for and make use of structure. MP8 Look for and express regularity in repeated reasoning.	TEKS-M 1.D Communicate mathematical ideas, reasoning, and their implications using multiple representations, including symbols, diagrams, graphs, and language as appropriate.

Elementary Ideas

- Discuss the importance of concisely and efficiently sketching mathematical ideas when making sense of problems.
- Have students practice with base-ten blocks by building a number and sketching it. Students should use large squares, lines, and dots.

$$200 \ + \ 30 \ + \ 6$$

- Have students sketch diagrams of word problems to show their comprehension of the context.

MP7 Look for and make use of structure. MP8 Look for and express regularity in repeated reasoning.	TEKS-M 1.D Communicate mathematical ideas, reasoning, and their implications using multiple representations, including symbols, diagrams, graphs, and language as appropriate.

Secondary Ideas

- Discuss the importance of concisely and efficiently sketching mathematical ideas when making sense of problems.

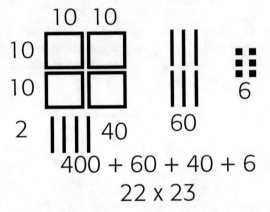

$$400 + 60 + 40 + 6$$
$$22 \times 23$$

- Have students use diagrams to show how algorithms work.
- Have students sketch diagrams of word problems to show comprehension of the context before calculating.

Day 11: Self-Monitoring and Becoming an Autonomous Learner

On Day 11, model how to set learning targets with a statement about what students want to accomplish, such as "I can keep working to solve a problem, even if it seems hard" or "I can make a mistake and learn from it to persevere and make connections with things I already know." Make mathematical practices transparent and that involve students self-evaluating their developing habits of mind as mathematicians. Introduce mathematical practice and process standards MP3 and MP7 or TEKS-M 1.F.

MP3 Construct viable arguments and critique the reasoning of others. **MP7 Look for and make use of structure.**	**TEKS-M 1.F Analyze mathematical relationships to connect and communicate mathematical ideas.**

Elementary Ideas

- Discuss what it means to learn in a student-centered mathematics classroom. Encourage students to take risks by engaging in group discussions, reflective writing in math journals, and problem-based learning.

- Have students practice with Daily Mystery Number routines. Post a number and have students work together to think about all the ways they can represent that number. Have them make posters and hold a gallery walk to share, compare, and learn from each other. Point out that the learning was student-centered.

- Use one of Graham Fletcher's 3 Acts problems (see gfletchy.com/3-act). Hold a mathematical discussion after several days of exploring a problem act-by-act. Student work should show how mathematicians ask questions, persevere over time, and see patterns and relationships to solve problems.

Secondary Ideas

- Discuss what it means to learn in a student-centered mathematics classroom. Share the importance of taking risks by engaging in group discussions, reflective writing in math journals, and problem-based learning. Students should take an active part in the group.

- Have students practice with Daily Mystery Number routines. Post a number and have students work together to think about all the ways they can represent that number. Have them make posters and hold a gallery walk to share, compare, and learn from each other. Point out that the learning was student-centered.

- Pose the *5-Digit Magic Number Lesson Plan* on pages 180–181 to students. Ask them to work together to determine whether it is a magic trick or a mathematical problem. Encourage them to pose questions to find patterns and relationships in this seemingly magical problem.

Day 12: Justification through Genuine Questions

Mathematicians justify their thinking by sharing orally and in writing. Have students share accurate and appropriate evidence of their claims, including references to the text or prior classroom and real-life experiences. When students understand it is their responsibility to justify their thinking and solutions, they deepen their understanding of mathematical concepts.

Have students share models and examples of quality work to justify their reasoning, such as journal entries and student work. Remind students that *justification* means "explaining, describing, proving, reasoning, and showing understanding." Introduce mathematical practice and process standards MP2 and MP3 or TEKS-M 1.F.

MP2 Reason abstractly and quantitatively. **MP3 Construct viable arguments and critique the reasoning of others.**	**TEKS-M 1.F Analyze mathematical relationships to connect and communicate mathematical ideas.**

Elementary Ideas

- Introduce or review the Genuine Question posters in the room: *How can you prove that? Can you convince us?*
- Discuss the importance of questioning and not giving solutions. Have students use the Genuine Question posters when talking with each other. Posters will help them be a coach or give classmates think time.
- Hold a mental Math Talk using grade-level appropriate problems. For example, in kindergarten, post two ten frames and ask students to solve how many there are and share strategies (3 + 7). Then, solve 7 + 5 + 3 and 3 + 6 + 7. Ask students to tell what they noticed. Explain that there are often many ways to solve a problem. When teachers are coached or use think time, they show the math community that they value everyone's thinking.
- Solve one grade-level appropriate problem. Have students write how they solved the problem with words and pictures.

MP2 Reason abstractly and quantitatively.	TEKS-M 1.F Analyze mathematical relationships to connect and communicate mathematical ideas.
MP3 Construct viable arguments and critique the reasoning of others.	

Secondary Ideas

- Introduce or review the Genuine Question posters in the room: *How can you prove that? Can you convince us?* Brainstorm with students various situations where each question may be most useful. It is important to have students think about when to use which questions to further their mathematical investigations.

- Hold a Math Talk using grade-level appropriate problems, such as $6.3 - 2.7$ or $9.8 + 8.7$. Discuss the importance of questioning and not shouting solutions. Share with students that they can use the Genuine Question posters when talking with each other.

- Discuss that there are often many ways to solve a problem. Explain that when teachers are coached or use think time, they show the math community that they value everyone's thinking.

- Solve one grade-level appropriate problem. Have students write how they solved the problem with words and pictures.

Day 13: Asking Clarifying and Challenging Questions

Student questioning is essential to a lively, thoughtful mathematical discussion. Without students posing questions, there can be no shared inquiry or evidence of student curiosity. On Day 13, discuss how students can direct their questions to deepen understanding and show curiosity about mathematical concepts or problems presented. Introduce mathematical practice and process standards MP3 and MP6 or TEKS-M 1.F.

Role-play how to question each other while students listen to and seek understanding of ideas different from their own:

- Ask questions when curious.

- Ask questions about assumptions or thinking.

- Pose questions to clarify thinking or reasoning behind an argument or conclusion.

- Ask questions to clarify and understand the substance of the mathematics or speaker's point of view.

- Ask *what if* questions to encourage and challenge thinking.

- Use silence after another student stops speaking to provide think time. During think time, reflect about what was shared.

- Accurately paraphrase what another student is thinking or saying.

MP3 Construct viable arguments and critique the reasoning of others. **MP6 Attend to precision.**	**TEKS-M 1.F Analyze mathematical relationships to connect and communicate mathematical ideas.**

Elementary Ideas

- Have students ask each other questions that clarify other students' thinking, such as "Why did you put the 4 there?" or "How did you know it was 5?"

- Role-play how to ask questions that challenge students to deeply reflect on their own ideas, such as "How do you know what you know?" or "Can you clarify the steps in your thinking?" or "Why did you....?"

- Show students a common misconception, such as $6 + 3 = 10$ or $11 \times 11 = 111$. Have them think about the solution, and ask "Is the problem correct, or not?", "How might I justify my thinking?", or "What clarifying or challenging questions would I pose?"

MP3 Construct viable arguments and critique the reasoning of others. **MP6 Attend to precision.**	**TEKS-M 1.F Analyze mathematical relationships to connect and communicate mathematical ideas.**

Secondary Ideas

- Have students ask each other questions that clarify other students' thinking, such as "Why did you put the 4 there?" or "How did you know it was 5?"

- Role-play how to ask questions that will challenge students to deeply reflect on their own ideas, such as "How do you know what you know?" or "Can you clarify the steps in your thinking?" or "Why did you…?"

- Show students a common misconception, such as $23 \times 26 = 418$. Have them think about the solution, and ask "Is the problem correct, or not?", "How might I justify my thinking?" or "What clarifying or challenging questions would I pose?"

Day 14: The Power of Mistakes or Partial Thinking

Discuss how making mistakes and learning from them is powerful. When mathematicians make mistakes, they stop and listen to others' ideas, engage in discourse, and eventually solve problems. When mathematics is thought of in this way, as a creative learning process to make connections, learning, growth, and mistakes are encouraged and accepted as learning opportunities. Then, amazing things happen.

Share examples of mistakes that created mathematical successes with students, and explain how to stay positive when making mistakes. Introduce mathematical practice and process standards MP1 and MP2 or TEKS-M 1.C.

MP1 Make sense of problems and persevere in solving them. MP2 Reason abstractly and quantitatively.	TEKS-M 1.C Select tools, including real objects, manipulatives, paper and pencil, and technology as appropriate, and techniques, including mental math, estimate, and number sense as appropriate, to solve problems.

Elementary Ideas

- Share Jo Boaler's video about mistakes. https://www.youtube.com/watch?v=exmCR28kmZk
- Role-play what to say and how to act when a mistake is made.
- Read selections from the book *Mistakes That Worked: The World's Familiar Inventions and How They Came to Be* by Charlotte Foltz Jones.

Secondary Ideas

- Share Jo Boaler's video about mistakes. https://www.youtube.com/watch?v=exmCR28kmZk
- Share the video "My Favorite No" with students. Ask them to discuss how they would feel if their mistake was chosen? Why might this be considered a powerful learning activity? https://www.teachingchannel.org/videos/class-warm-up-routine
- Read selections from the book *Mistakes That Worked: The World's Familiar Inventions and How They Came to Be* by Charlotte Foltz Jones.

Day 15: Productive versus Non-Productive Noise

Explain to students that the old way of teaching was simple. Teachers used to stand in front of the room and lecture or tell students what to do. Tell students that the classroom will be busy, interactive, and flexible, and they may be up front as much as the teacher. Affirm that students should expect productive noise. Explain that productive noise means that students are talking like mathematicians, sharing mathematical strategies, and exploring and making sense of mathematics as a team. Introduce mathematical practice and process standards MP3 and MP6 or TEKS-M 1.E.

MP3 Construct viable arguments and critique the reasoning of others. MP6 Attend to precision.	TEKS-M 1.E Create and use representations to organize, record, and communicate mathematical ideas.

Elementary Ideas

- Create a chart that describes what productive noise sounds like versus what unproductive noise sounds like.
- With permission, videotape students working in pairs or small groups. Have the whole class reflect on the level of productive noise in the recording and how it should look and sound.
- Show a video clip of a classroom where students are engaged in productive noise. Record what students noticed on chart paper.

Secondary Ideas

- Create a chart that describes what productive noise sounds like versus what unproductive noise sounds like.
- Discuss how students feel about noise when they puzzle and think about mathematical concepts.
- Show a video clip of a classroom where students are engaged in productive noise. Record what students noticed on chart paper.

Day 16: Being a Problem Solver

Teaching students how to be independent problem solvers means explicitly teaching the process. Discuss what it means to persevere and what processes and strategies mathematicians use to solve problems. Introduce mathematical practice and process standards MP1 and MP6 or TEKS-M G.1. Share George Polya's four steps for doing mathematics (1948):

1. **Understanding the problem.** Have students figure out what the problem is about and identify the question posed.

2. **Devising a plan.** Allow students time to think about how to solve the problem: *Do I want to use tools, models, or procedures?*

3. **Carrying out the plan.** Have students implement their plans based on understanding of the context and mathematics of the problem.

4. **Looking back.** Have students reflect: *Does my thinking from Step 3 answer the problem as originally understood in Step 1? Does my answer make sense?*

MP1 Make sense of problems and persevere in solving them. MP6 Attend to precision.	TEKS-M 1.G Display, explain, and justify mathematical ideas and arguments using precise mathematical language in written or oral communication.

Elementary Ideas

- Play the game Four Corners. (Note: There is no correct answer.) Place one of the following problem-solving statements in each corner of the room: *problem solving requires patience, problem solving requires persistence, problem solving requires risk-taking,* and *problem solving requires cooperation.*

  - Ask students to choose the statement most true at this point and silently walk to that corner. Have students discuss why they made their choices. Call on one student in each group to share. Then, have a class discussion about what it means to be a problem solver using Polya's four steps.

- Discuss strategies if students get stuck in the problem-solving process: restate the problem, use a tool or manipulative, talk with a partner about the problem, try a different strategy, and think about what is known in the problem.

- Create a math problem together that all can solve in small groups, such as *There were _____ (number from 5 to 25) _____ (kind of animal or vehicle). How many _____? (parts) were there?*

- Use problems from *What's Your Math Problem!?! Getting to the Heart of Teaching Problem Solving* or *50 Leveled Math Problems for Your Grade Level.*

MP1 Make sense of problems and persevere in solving them. MP6 Attend to precision.	TEKS-M 1.G Display, explain, and justify mathematical ideas and arguments using precise mathematical language in written or oral communication.

Secondary Ideas

- Play the game Four Corners. Place one of the following problem-solving statements in each corner of the room: *problem solving requires patience*, *problem solving requires persistence*, *problem solving requires risk-taking*, and *problem solving requires cooperation*.

  - Ask students to choose the statement most true at this point and silently walk to that corner. Have students discuss why they made their choices. Call on one student in each group to share. Then, have a class discussion about what it means to be a problem solver using Polya's four steps.

- Discuss the idea that many times we are told how to do math. Discuss why it might be important to struggle and puzzle over math rather than just find a solution. Ask students how this might help them in the future.

- Provide problems from Robert Kaplinsky's website robertkaplinsky.com/lessons/. Why are they fun to work hard on?

- Discuss some of the strategies students can use if they get stuck in the problem-solving process: restate the problem in your own words, use a tool or manipulative, talk with a friend about the problem, jot down ideas or draw a diagram about what you do know, try a different strategy, think about what you do know about the information in the problem, take a break for a few minutes, and look back in your journal for similar problems.

- Use problems from *What's Your Math Problem!?!: Getting to the Heart of Teaching Problem Solving* or *50 Leveled Math Problems for Your Grade Level*.

Day 17: Using the Three Reads Problem-Solving Strategy

On Day 17, discuss what problem solving is and what mathematical problem solvers need to think about. Explain that, to problem solve, the context must be identified and mathematical quantities and relationships defined. Have students explore how sketching a diagram of mathematical relationships and contexts poses different questions. Questions can be answered through this reading process. Use the Three Reads problem-solving strategy with a mathematics concept that has already been taught to focus the lesson on the structure of the strategy. Let students know that this structure will be used to solve problems throughout the year. Review community agreements and help students know that they need to respect everyone's ideas. Introduce mathematical practice and process standards MP1 and MP8 or TEKS-M 1.B.

MP1 Make sense of problems and persevere in solving them. **MP8 Look for and express regularity in repeated reasoning.**	**TEKS-M 1.B Use a problem-solving model that incorporates analyzing given information, formulating a plan or strategy, determining a solution, justifying the solution, and evaluating the problem-solving process and the reasonableness of the solution.**

Elementary Ideas

- Have students solve one of the following problem stems using the Three Reads problem-solving strategy:
  - Ten of us went to the snow during winter break. We wanted to make snowmen. Each snowman needed 1 carrot for a nose, 2 pinecones for eyes, and 3 pieces of coal for buttons.
  - Ellie used 4 lemons to make a 3-liter pitcher of lemonade. She needs to make 12 liters for her family; 24 liters for her girl scout troop; and 36 liters for her birthday party.
  - Hannah's age this year is a multiple of 5. Next year Hannah's age will be a multiple of 4.
  - Kimberly made a plate of cookies. She had the same number of chocolate chip cookies as oatmeal cookies.

MP1 Make sense of problems and persevere in solving them. **MP8 Look for and express regularity in repeated reasoning.**	**TEKS-M 1.B Use a problem-solving model that incorporates analyzing given information, formulating a plan or strategy, determining a solution, justifying the solution, and evaluating the problem-solving process and the reasonableness of the solution.**

Secondary Ideas

- Have students solve one of the following problems using the Three Reads problem-solving strategy:
  - Together Jackson, Daisy, and Lena had $865 dollars when they started shopping. Jackson spent $\frac{2}{5}$ of his money. Daisy spent $40.00. Lena spent twice as much as Jackson. They each have the same amount of money left over.
  - Jessica is making fruit smoothies for her friends using raspberries, blackberries, blueberries, and cherries. She needs a total of 280 pieces of fruit. There are 2 times as many blackberries as raspberries, 3 times as many blueberries as cherries, and 4 times as many cherries as blackberries.
  - On Saturday morning, Susan and Becky each have the same amount of money. Saturday afternoon, Susan spends $166.00 buying clothes and Becky spends $139.00 buying a concert ticket. On Saturday night, the ratio of Susan's money to Becky's money is 1:4.

Day 18: Games in Math—Your Role as a Mathematician

Students of all ages love to play fun and motivating games. Tell students they will engage in game-like math activities throughout the school year. Explain that games are a great way for them to deepen mathematical understanding and reasoning. Some of the reasons to play games this year are that they:

- Provide meaningful practice;

- Help students develop fluency when played repeatedly;

- Encourage strategic mathematical thinking;

- Present opportunities to engage in mathematics without the teacher present;

- Give the teacher opportunities to differentiate instruction; and

- Give the teacher opportunities to convene small groups to support or challenge students with mathematical content.

On Day 18, discuss the norms expected when students play games during math activities or math learning center times. Explain that students will play games together and that everyone will know the rules and talk about the mathematics in the game. Introduce mathematical practice and process standards MP4 and MP5 or TEKS-M 1.C. It is important to have a well-organized system that puts students in charge of the activity time.

MP4 Model with mathematics. MP5 Use appropriate tools strategically.	TEKS-M 1.C Select tools, including real objects, manipulatives, paper and pencil, and technology as appropriate, and techniques, including mental math, estimation, and number sense as appropriate, to solve problems.

Elementary Ideas

- Introduce some of your favorite apps or websites for students to explore. How can they use these to deepen their mathematical thinking and problem solving? See www.mathlearningcenter.org/apps for some free apps.
- Use a Daily Math Stretch to build conceptual understanding in a game-like warm-up.
- Consider using other resources, such as books like *Well Played: Building Mathematical Thinking Through Number Games and Puzzles K–2* or *3–5*.

MP4 Model with mathematics. MP5 Use appropriate tools strategically.	TEKS-M 1.C Select tools, including real objects, manipulatives, paper and pencil, and technology as appropriate, and techniques, including mental math, estimation, and number sense as appropriate, to solve problems.

Secondary Ideas

- Introduce some of your favorite apps or websites for students to explore. How can they use these to deepen their mathematical thinking and problem solving?
- Play favorite games that explore math concepts at your grade level.
- Pose a math puzzle and let students engage in discourse about the mathematics in the activity that may seem somewhat magical. For example, 5-Digit Number Magic Problem on pages 180–181. Discuss how the game reflects the definition of mathematics (i.e., the constant search for patterns and relationships).
- Use a Daily Math Stretch to build conceptual understanding in a game-like warm-up in *Daily Math Stretches: Building Conceptual Understanding*.
- Consider using other resources, such as books like *Well Played: Building Mathematical Thinking Through Number Games and Puzzles 6–8*.

Day 19: Small Group and Workstation Discourse

On Day 19, focus on small group and workstation discourse. Let students know that small groups give teachers time to work with students on specific concepts. They might work on reasoning, logic, basic math skills, and strategies to solve problems. During small group instruction, the rest of the class may work independently at workstations.

Math discourse can fit well into small group and workstation instruction. Instructional frameworks, such as Guided Math, that include these settings are particularly nuanced for math discourse because they share similar principles promoting exploration and risk taking and supporting students to see themselves as mathematicians. (For more information on the Guided Math framework, see Laney Sammons' books: *Guided Math: A Framework for Mathematics Instruction, Strategies for Implementing Guided Math,* and *Guided Math Workstations.*)

To implement discourse with small group instruction, place 3–6 students in a group and scaffold various levels of understanding while using new strategies, models, and materials. Plan 5–10 minute sessions that support or extend mathematical concepts taught during the whole group lesson. Then, students may work on similar activities independently at workstations. Introduce mathematical practice and process standards MP1 and MP7 or TEKS-M 1.B.

MP1 Make sense of problems and persevere in solving them. **MP7 Look for and make use of structure.**	**TEKS-M 1.B Use a problem-solving model that incorporates analyzing given information, formulating a plan or strategy, determining a solution, justifying the solution, and evaluating the problem-solving process and the reasonableness of the solution.**

Elementary Ideas

- Plan a small group lesson using a problem-solving approach.
- Discuss routines to make sure students are autonomous and do not interrupt small group work.

Secondary Ideas

- Plan a small group lesson using a problem-solving approach.
- Discuss routines to make sure students are autonomous and do not interrupt small group work.

Day 20: Talk Moves to Create Good Mathematical Conversation without the Teacher!

On Day 20, revisit your community agreements and create any still needed. Review the important talk moves that all students have used to engage in mathematical discourse. It will be important to remind students that math discourse is not telling others what the answer is. Talk should be everyone discussing a problem, whether it be partial thinking or a solution. Encourage students to puzzle and fight for sensemaking when discussing mathematical situations.

Remind students about other agreements that have been made, such as waiting for others, ask groups member to add onto one another's ideas and taking time to reflect and reason about what is being discussed. Reflection might include, *Do I agree or disagree with that idea? and Why?* Review the importance of re-voicing others' ideas by using language, such as "So, you are saying that _____. Do I have that right?" If there is time, review routines and protocols that have helped students, and remind them that they can use routines and protocols anytime in mathematical discussions. Emphasize that this is what being an autonomous learner and mathematician feels like!

Discuss how peer scaffolding (e.g., Think-Pair-Share) will not actively be part of the building of knowledge. Have this discussion early to deepen mathematical discourse. Introduce mathematical practice and process standards MP3 and MP6 or TEKS-M 1.G.

MP3 Construct viable arguments and critique the reasoning of others. **MP6 Attend to precision.**	**TEKS-M 1.G Display, explain, and justify mathematical ideas and arguments using precise mathematical language in written or oral communication.**

Elementary Ideas

- Read one of the following books to spark a conversation about discourse and what it means to persevere and fight for sensemaking.
  - *The Man Who Walked Between the Towers* by Mordicai Gerstein
  - *Thank You, Mr. Falker* by Patricia Polacco
  - *Amazing Grace* by Mary Hoffman
  - *Mistakes That Worked: The World's Familiar Inventions and How They Came to Be* by Charlotte Foltz Jones

MP3 Construct viable arguments and critique the reasoning of others. **MP6 Attend to precision.**	**TEKS-M 1.G Display, explain, and justify mathematical ideas and arguments using precise mathematical language in written or oral communication.**

Secondary Ideas

- Read one of the following books to spark a conversation about discourse and what it means to persevere and fight for sensemaking.

  - *Winterdance: The Fine Madness* by Gary Paulsen

  - *Mistakes That Worked: The World's Familiar Inventions and How They Came to Be* by Charlotte Foltz Jones

- Play Four Corners again by placing these four statements in each corner and ask students to stand by a statement that resonates with them right now.

  - It is easier to persevere when I work in cooperation with my peers.

  - When I am stuck, it helps when I turn the problem into a picture or diagram.

  - It is easier to persevere when the problem is like a game.

  - I keep persevering when I am praised for working hard instead of being told I am correct.

Conclusion

Taking the time to let students talk, make sense of the mathematics, and puzzle with excellent tasks creates opportunities for all to become autonomous learners. Beginning this process using the First 20 Days of Discourse will change the environment of your math classroom and deepen your students' talk, inquiry, and problem-solving skills.

As you practice asking genuine questions to assess and scaffold, student understanding will become a natural pedagogy. You will not remember a time when you didn't facilitate discourse in your mathematics classroom. You will see students learning to use visual models and representations to support their thinking. You will be able to see firsthand students' thinking and use it to modify and enhance learning opportunities on a daily basis.

Teachers who take the time to plan and implement effective instructional routines described in this book will be amazed at the depth of student learning and academic growth they see. Through the use of discourse, teachers will be able to integrate and structure the cultural and linguistic resources unique to their students to create learning environments that extend the mathematical learning for all, ensuring learning is concerned with each student's sense of mathematical identity.

Final Thoughts

As you use this resource throughout the year, focus on the following teacher actions:

- Consistently focus lessons on the mathematical practice and process standards to build habits of mind of a mathematician for all students.

- Elicit, value, and celebrate varied approaches and solution paths students use to explore and solve mathematical problems, and learn from your students.

- Promote talk, explanations, and challenging questions that critique mathematical reasoning and provide opportunities for all students to learn from each other.

- Plan for and teach lessons that show the beauty of learning mathematics by promoting curiosity, self-confidence, flexibility, and perseverance.

- Model high expectations for every student's success in problem solving, reasoning, and making sense of the mathematics.

- Promote a growth mindset among students.

- Focus on connections that relate key math ideas to real-world and mathematical contexts.

- Incorporate mathematical tools as an everyday part of the mathematics classroom.

- Continue to grow in your knowledge of mathematics for teaching, and strive to be a lifelong learner.

Remember, only when these words become actions will we see more productive beliefs, new norms, and agreements about what instructional practice should look and sound like, and begin to ensure mathematical success.

Reflect and Discuss

1. How will you talk with students about building a mathematical community of learners?

2. What are your expectations for accountable talk and active listening?

3. How will you encourage students to use genuine questions, justification, and evidence this year?

4. What will you communicate when students make mistakes or are unsure about the math they use?

5. How will letting students talk more change the structure of your instruction?

References Cited

Arturias, Harold, and Phil Daro. 2013. "Moving Beyond the Answer." https://www.aimsedu.org/2013/02/01/moving-beyond-the-answer/.

Ball, Deborah Lowenberg. 1991. "What's All This Talk about 'Discourse'?" *Arithmetic Teacher* 39 (3): 44–48.

Ball, Deborah Loewenberg, Mark Hoover Thames, and Geoffrey Phelps. 2008. "Content Knowledge for Teaching: What Makes It Special?" *Journal of Teacher Education* 59 (5): 389–407.

Bidwell, James K., and Robert G. Clason, eds. 1970. *Readings in the History of Mathematics Education.* Washington DC: NCTM.

Blanke, Barbara. 2009. "Understanding Mathematical Discourse in the Elementary Classroom: A Case Study." PhD diss., Oregon State University.

Boaler, Jo. 2008. *What's Math Got to Do with It?: Helping Children Learn to Love Their Least Favorite Subject and Why It's Important for America.* New York: Viking Adult.

——. 2016. *Mathematical Mindsets: Unleashing Students' Potential through Creative Math, Inspiring Messages and Innovative Teaching.* San Francisco, CA: Jossey-Bass.

——. 2017. "Setting Up Positive Norms in Math Class." Youcubed. https://bhi61nm2cr3mkdgk1dtaov18-wpengine.netdna-ssl.com/wp-content/uploads/2017/03/Norms-Paper-2015.pdf.

Boaler, Jo, and Cathleen Humphreys. 2005. *Connecting Mathematical Ideas: Middle School Video Cases to Support Teaching and Learning.* Portsmouth, NH: Heinneman.

Boaler, Jo, and Karin Brodie. 2004. "The Importance, Nature, and Impact of Teacher Questions." In *Proceedings of the 26th Conference of the Psychology of Mathematics Education* edited by Douglas E. McDougall and John A. Ross, North America, 773–781. Toronto: OISE/UT.

Boring, Edwin G. 1957. *A History of Experimental Psychology*. 2nd ed. New York: Appleton-Century-Crofts.

Bresser, Rusty, Kathy Melanese, and Christine Sphar. 2009. *Supporting English Language Learners in Math Class.* Sausalito, CA: Math Solutions.

Britannica. 2001. "No Child Left Behind." https://www.britannica.com/topic/No-Child-Left-Behind-Act.

Brooks, Edward. 1880. *The Philosophy of Arithmetic: The Philosophy of Arithmetic as Developed From the Three Fundamental Processes of Synthesis, Analysis and Comparison, Containing Also a History of Arithmetic*. Lancaster, PA: Normal Publishing Co.

Bruner, Jerome S. 1960. *The Process of Education.* Cambridge, MA: Harvard University Press.

Burris, Carol Corbett, Jay P. Heubert, Henry M. Levin. 2006. "Accelerating Mathematics Achievement Using Heterogeneous Grouping." *American Educational Research* 43 (1): 103–104.

Carpenter, Thomas P., Elizabeth Fennema, Megan Loef L. Franke, Linda Levi, and Susan B. Empson. 2015. *Children's Mathematics: Cognitively Guided Instruction.* 2nd edition. Portsmouth, NH: Heinemann/Reston, VA: NCTM.

Carter, Susan. 2008. "Disequilibrium and Questioning in the Primary Classroom: Establishing Routines That Help Students Learn." *Teaching Children Mathematics* 3 (15): 134–137.

Chapin, Suzanne, and Kristin Eastman. 1996. "External and Internal Characteristics of Learning Environments (Implementing the Professional Standards for Teaching Mathematics Department)." *Mathematics Teacher* 89 (2): 112–115.

Chapin, Suzannne H., Catherine O'Connor, and Nancy Canavan Anderson. 2009. *Classroom Discussion in Math: A Teacher's Guide for Using Talk Moves to Support the Common Core and More*. Sausalito, CA: Math Solutions.

Cobb, Paul. 2007. "Putting Philosophy to Work: Coping with Multiple Theoretical Perspectives." In *Second Handbook for Research on Mathematics Teaching and Learning* edited by Frank K. Lester Jr. 3–38. Charlotte, NC: Information Age Publishing, Inc.

Colburn, Warren. 1849. *Colburn's First Lessons: Intellectual Arithmetic upon the Inductive Method of Instruction.* Boston, MA: William J. Reynolds.

Darling-Hammond, Linda. 2000. *Teacher Quality and Student Achievement.* Vol. 8 no. 1. Seattle: Education Analysis Archives.

Daro, Phil. 2017. "Moving Beyond Answer-getting." http://math.serpmedia.org/sensemaking/triangles.html.

Davies, Charles. 1850. *The Logic and Utility of Mathematics with the Best Methods of Instruction Explained and Illustrated.* New York: A. S. Barnes & Co.

Davis, Robert B., Carolyn C. Maher, and Nel Noddings. 1990. *Constructivist View on the Teaching and Learning of Mathematics.* Monograph Series No. 4. 1–3, 195–210. Reston, VA: NCTM.

DeVault, M. Vere, and J. Fred Weaver. 1970. "Forces and Issues Related to Curriculum and Instruction, K–6." In *A History of Mathematics Education in the United States and Canada.* Vol. 32, edited by P.S. Jones and J. Coxford, A.F. 93–152 Washington DC: NCTM.

Dewey, John. 1916. *Democracy and Education: An Introduction to the Philosophy of Education.* New York: MacMillian.

Donovan, M. Suzanne, and John D. Bransford, eds. 2005. *How Students Learn: History, Mathematics, and Science in the Classroom.* National Research Council, Committee on How People Learn: A Targeted Report for Teachers. Washington, DC: National Academies Press.

Duckworth, Eleanor. 1987. *"The Having of Wonderful Ideas" and Other Essays on Teaching and Learning.* New York: Teacher's College Press.

Dweck, Carol S. 2006. *Mindset: The New Psychology of Success.* New York: Ballantine Books.

——. 2015. "Carol Dweck Revisits the 'Growth Mindset'." *Education Week* 35 (5): 20, 24.

———. 2016. *Mindset: The New Psychology of Success: How We Can Learn to Fulfill Our Potential.* NY: Random House, 16.

Foreman, Linda Cooper, and Albert B. Bennett Jr. 1996. "Paperfolding." *Visual Mathematics, Course II.* Salem, OR: The Math Learning Center.

Fosnot, Catherine Twomey. 2005. *Constructivism: Theory, Perspectives, and Practice.* 2nd ed. NY: Teachers College Press.

Froehle, Craig. 2012. "Engagement and Equity."

Harris Weber, Pamela. 2011. *Building (Powerful) Numeracy for Middle and High School Students.* Portsmouth, NH: Heinemann.

———. 2014. *Lessons and Activities For Building Powerful Numeracy.* Portsmouth, NH: Heinemann.

Herbel-Eisenmann, Beth, and Michelle Cirillo, eds. 2009. *Promoting Purposeful Discourse: Teacher Research in Mathematics Classrooms.* Reston, VA: NCTM.

Higgins, Karen, Cary Cermak-Rudolf, Barbara Blanke. 2009. "Yeah, But What If…? A Study of Mathematical Discourse in a Third Grade Classroom." In *The Role of Mathematics Discourse in Producing Leaders of Discourse,* edited by Libby Knott. The Montana Mathematics Enthusiast Monograph Series in Mathematics Education.

Hollie, Sharroky. 2015. *Strategies for Culturally and Linguistically Responsive Teaching and Learning: Building Receptive Language for All Students.* Huntington Beach, CA: Shell Educational Publishing.

Hughes, John, dir. 1986. *Ferris Bueller's Day Off.* Los Angeles, CA: Paramount Pictures.

Humphreys, Cathy, and Ruth Parker. 2015. *Making Number Talks Matter.* Portland, ME: Stenhouse Publishers.

James, William. 1957. *The Principles of Psychology.* New York: Henry Holt and Company.

Kelemanik, Grace, Amy Lucenta, and Susan Janssen Creighton. 2016. *Routines for Reasoning: Fostering the Mathematical Practices in All Students.* Portsmouth, NH: Heinemann.

Lambdin, Diana V., and Crystal Walcott. 2007. "Change through the Years: Connections between Psychological Learning Theories and the School Mathematics Curriculum." In *The Learning of Mathematics, 69 Yearbook,* edited by W. Gary Martin, Marilyn E. Strutchens, Portia Elliot. 3–25. Reston, VA: NCTM.

Leinwand, Steve. 2016. foreword to *Number Talks: Fractions, Decimals, and Percentages* by Sherry Parrish and Ann Dominick. Sausalito, CA: Math Solutions.

Ma, Liping. 1999. *Knowing and Teaching Elementary Mathematics.* NY: Routledge.

Math Learning Center, The. 2015a. "Math Practice Posters." mathlearningcenter.org

——. 2015b. "Multiplication Fact Strategy Posters." mathlearningcenter.org

McCallum, Bill. 2011. "Structuring the Mathematical Practices." (blog) *Mathematical Musings.* http://mathematicalmusings.org/2011/03/10/structuring-the-mathematical-practices/.

Mehan, Hugh. 1979. *Learning Lessons: Social Organization in the Classroom.* Cambridge, MA: Harvard University Press.

Mora-Flores, Eugenia, and Angelica Machado. 2015. *Strategies for Connecting Content and Language for English Language Learners in Mathematics.* Huntington Beach, CA: Shell Educational Publishing.

National Council of Teachers of Mathematics (NCTM). 1935. *Teaching of Arithmetic: 10 Yearbook,* edited by W.D. Reeve. 1–31. NY: Bureau of Publications, Teachers College, Columbia University.

——. 1989. *Curriculum and Evaluation Standards for School Mathematics.* NCTM: Reston, VA.

——. 2000. *Principles and Standards for School Mathematics.* NCTM: Reston, VA.

——. 2005. *How Students Learn: History, Mathematics, and Science in the Classroom.* Washington DC: National Academy Press.

——. 2007. *Mathematics Teaching Today: Improving Practice, Improving Student Learning.* 2nd ed. revised version 1991, edited by Tami S. Mark, Reston. VA: NCTM.

——. 2009. *Focus in High School Mathematics: Reasoning and Sense Making.* Reston, VA: NCTM.

———. 2014. *Principles to Actions: Ensuring Mathematical Success for All.* Reston, VA: NCTM.

National Governors Association (NGA) Center for Best Practices and Council of Chief State School Officers (CCSSO). 2010. *Common Core State Standards: English Language Arts.* Washington, DC: National Governors Association Center for Best Practices, Council of Chief State School Officers. www.corestandards.org.

———. 2010. *Common Core State Standards: Mathematics.* Washington, DC: National Governors Association Center for Best Practices, Council of Chief State School Officers. www.corestandards.org.

National Research Council. 1996. *National Science Education Standards.* Washington, DC: National Academy Press.

———. 2001. *Adding It Up: Helping Children Learn Mathematics.* Washington DC: National Academy Press.

———. 2005. *How Students Learn: History, Mathematics, and Science in the Classroom.* 13. Washington DC: the National Academies Press.

———. 2012. *Education for Life and Work: Developing Transferable Knowledge and Skills in the 21st Century.* James W. Pellegrino and Margaret L. Hilton, eds. Committee on Defining Deeper Learning and 21st Century Skills, Board on Testing and Assessment and Board on Science Education, Division of Behavioral and Social Sciences and Education. Washington, DC: National Academies Press.

Network Communicate Support Motivate (NCSM). 2011. Summer Leadership Academy. Conference held at Atlanta, GA.

O'Connell, Susan. 2016. *Math in Practice: A Guide for Teachers.* Portsmouth, NH: Heinemann.

Parrish, Sherry. 2010. *Number Talks: Helping Children Build Mental Math and Computations Strategies.* Sausalito, CA: Math Solutions.

Pasachoff, Naomi. 2017. "Marie Curie: Her Story in Brief." American Institute of Physics. https://history.aip.org/history/exhibits/curie/brief/index.html.

Piaget, Jean. 1952. *The Child's Conception of Number.* New York: Humanities Press.

———. 1970a. *Piaget's Theory.* New York: Wiley.

——. 1970b. *Science of Education and the Psychology of the Child.* New York: Orion Press.

Pike, Nicolas. 1788. *New and Complete System of Arithmetic—Composed for the Use of the Citizens of the United States.* Worcester, MA.

Polya, George. 1948. *How to Solve It: A New Aspect of Mathematical Method.* Reprinted in 2009. Japan: Ishi Press.

Ronda, Erlina. 2012. "The Four Freedoms in the Classroom." Mathematics for Teaching. math4teaching.com/the-four-freedoms-in-the-classroom.

Russell, Susan Jo. 2000. "Developing Computational Fluency with Whole Numbers." *Teaching Children Mathematics* 7 (3): 154–158.

Schon, Donald A. 1987. *Educating the Reflective Practitioner: Toward a New Design for Teaching and Learning in the Professions.* San Francisco: Jossey-Bass.

Sherin, Miriam Gamoran. 2002. "A Balancing Act: Developing a Discourse Community in a Mathematics Classroom." *Journal of Mathematics Teacher Education* 5: 205–233.

Smith, John P. 1996. "Efficacy and Teaching Mathematics by Telling: A Challenge for Reform." *Journal for Research in Mathematics Education* 27 (4): 387–402.

Schon, Donald A. 1987. *Educating the Reflective Practitioner: Toward a New Design for Teaching and Learning in the Professions.* San Francisco, CA: Jossey-Bass.

Small, Marian. 2012. *Good Questions: Great Ways to Differentiate Mathematics Instruction.* 2nd edition. Columbia University: Teacher's College Press.

Smith, Margaret S., Elizabeth K. Hughes, Randi A. Engle, Mary Kay Stein. 2009. "Orchestrating Discussions." *Mathematics in the Middle School* 14 (9).

Smith, Margaret S., and Mary K. Stein. 1998. "Selecting and Creating Mathematical Tasks: From Research to Practice." *Mathematics Teaching in the Middle School* 3: 344–50.

——. 2011. *5 Practices for Orchestrating Productive Mathematics Discussions.* Reston, VA: NCTM/Thousand Oaks, CA: Corwin Press.

Stein, Mary K., Margaret S. Smith, Marjorie A. Henningsen, and Edward A. Silver. 2009. *Implementing Standards-Based Mathematics Instruction: A Casebook for Professional Development.* 2nd ed. New York: Teachers College Press.

Texas Education Agency. 2012. "Implementation of Texas Essential Knowledge and Skills for Mathematics, Elementary, Adopted 2012." Texas Essential Knowledge and Skills for Mathematics. http://ritter.tea.state.tx.us/rules/tac/chapter111/ch111a.html.

Thorndike, Edward Lee. 1922. *The Psychology of Arithmetic.* New York: MacMillan.

Treisman, Uri. 2016. Joint Mathematics Meeting. AMA and AMS. Washington State Convention Center, Seattle, WA.

Van de Walle, John A., Karen S. Karp, and Jennifer M. Bay-Williams. 2010. *Elementary and Middle School Mathematics: Teaching Developmentally,* 7th edition. New York: Allyn and Bacon.

Vygotsky, L. S. 1978. *Mind in Society: The Development of Higher Psychological Processes.* London: Harvard University Press.

——. 1986. *Thought and Language.* translated by Alex Kosulin. Cambridge, MA: Harvard University Press.

Weaver, Dave, and Tom Dick. 2006. "Assessing the Quality and Quantity of Student Discourse in Mathematics Classrooms Year 1 Results." http://ormath.mspnet.org/media/data/WeaverDick.pdf?media_000000006132.pdf.

Wood, Terry, Paul Cobb, and Erna Yackel. 1991. "Change in Teaching Mathematics: A Case Study." *American Educational Research Journal* 28 (3): 587–616.

Wood, Terry, and Tammy Turner-Vorbeck. 2001. "Extending the Conception of Mathematics Teaching." In *Beyond Classical Pedagogy: Teaching Elementary School Mathematics* edited by Terry Wood, Barbara Scott Nelson, and Janet Warfield. 185–208. Mahwah, NJ: Lawrence Erlbaum.

Yackel, Erna, and Paul Cobb. 1996. "Sociomathematical Norms, Argumentation, and Autonomy in Mathematics." *Journal for Research in Mathematics Education* 27 (4): 458–477.

Recommended Resources

Books

Boucher, Donna, and Laney Sammons. 2017. *Guided Math Workstations.* Huntington Beach, CA: Shell Educational Publishing.

Dacey, Linda, Karen Gartland, and Jayne Bamford Lynch. 2015. *Well Played: Building Mathematical Thinking Through Number Games and Puzzles.* Portland, ME: Stenhouse Publishers.

Gerstein, Mordicai. 2003. *The Man Who Walked Between the Towers.* New York: Square Fish.

Gojak, Linda. 2011. *What's Your Math Problem!?! Getting to the Heart of Teaching Problem Solving* or *50 Leveled Math Problems for Your Grade Level.* Huntington Beach, CA: Shell Educational Publishing.

Hoffman, Mary. 1991. *Amazing Grace.* New York: Dial Books.

Jones, Charlotte Foltz. 2016. *Mistakes That Worked: The World's Familiar Inventions and How They Came to Be.* New York: Delacorte Press.

Jones, Charlotte Foltz, and John O'Brien. 1991. *Mistakes That Worked: 40 Familiar Inventions & How They Came to Be.* New York: Delacorte Press.

McDonald P., Joseph, Nancy Mohr, Alan Dichter, and Elizabeth C. McDonald. 2013. *The Power of Protocols: An Educator's Guide to Better Practice, Third Edition.* New York: Teacher's College Press.

Paulsen, Gary. 1994. *Winterdance: The Fine Madness.* New York: Houghton Mifflin Harcourt Publishing Company.

Polacco, Patricia. 1998. *Thank You, Mr. Falker.* New York: Philomel Books.

Poyla, George. 1945. *How to Solve It.* La Vergne, TN: Lightning Source Inc.

Sammons, Laney. 2010. *Guided Math: A Framework for Mathematics Instruction.* Huntington Beach, CA: Shell Educational Publishing.

——. 2013. *Strategies for Implementing Guided Math.* Huntington Beach, CA: Shell Educational Publishing.

Websites

Achieve the Core
www.achievethecore.org

Database of Activity-Based Learning
bit.ly/MTBoSbank

Database of Math Activities in desmos
bit.ly/desmosbank

desmos
teacher.desmos.com

Divisible by 3 Andrew Stadel
mr-stadel.blogspot.com

Dr. Nicki's Guided Math Blog
guidedmath.wordpress.com/tag/daily-math-routines/

dy/dan from Dan Meyer
blog.mrmeyer.com

Estimation 180
www.estimation180.com

Geogebra Creations from Dr. Ted Coe
tedcoe.com/math

Gifsmos from Chris Lusto
www.gifsmos.com

Graham Fletcher's 3-Acts Lessons
gfletchy.com/3-act-.lessons

Illustrative Mathematics
www.illustrativemathematics.org

Inside Mathematics, "Number Talks"
www.insidemathematics.org/classroom-video/number-talks

Jo Boaler–Study Skills
www.youtube.com/watch?v=exmCR28kmZk

Jo Boaler, Youcubed
www.youcubed.org

Math Munch
mathmunch.org

Math Talks
www.mathtalks.net

Mathalicious, grades 6–12
www.mathalicious.com

Mathalicious Unit Plans
www.mathalicious.com/lessons

Mathematics Assessment Resource Service (MARS)
map.mathshell.org/materials/index.php

National Council of Teachers of Mathematics (NCTM)
www.nctm.org

NCTM Illuminations
illuminations.nctm.org

NCTM The Math Forum
www.mathforum.org

NRICH
nrich.maths.org

Number Strings
numberstrings.com

Problems of the Month—inside mathematics

www.insidemathematics.org/problems-of-the-month/download-problems-of-the-month

Robert Kaplinsky

robertkaplinsky.com/lessons/

Robert Kaplinsky Problem-Based Search Engine

bit.ly/RKsearchengine

San Diego City Schools, "Mathematics Routine Bank"

svmimac.org/images/Cristo_Rey_-_Middle_Level_Bank.pdf

San Francisco Unified School District, "More Resources for Math Talks"

www.sfusdmath.org/math-talks-resources.html

Strategies for Teaching Elementary Mathematics, "Mathematics Routines"

mathteachingstrategies.wordpress.com/2008/11/24/mathematical-routines/

Teaching Channel

www.teachingchannel.org/videos/class-warm-up-routine

The Classroom Chef

www.classroomchef.com/chefs

The Math Learning Center

www.mathlearningcenter.org

The Math Learning Center Apps

www.mathlearningcenter.org/apps

The Math Learning Center Math Vocabulary Cards

www.mathlearningcenter.org/resources/apps/math-vocabulary-cards

The MTBoS Search Engine

bit.ly/MTBOSS

The Reflective Educator, "Math Talk"

davidwees.com/content/math-talk

TMathC

tmathc.com/

Transformulas from Jed Butler

transformulas.org/

Visual Patterns: grades K–12

www.visualpatterns.org

Which One Doesn't Belong? (WODB)

wodb.ca

WODB Graphs and Equations

wodb.ca/graphs.html

WODB Numbers

wodb.ca/numbers.html

Yummy Math

www.yummymath.com/

5-Digit Magic Number Lesson Plan

Sample: 1st number (given by a student)　　23,795

2nd number (given by a student)　　64,028

3rd number (given by a teacher)　　35,971

4th number (given by a student)　　16,264

5th number (given by a teacher)　　<u>83,735</u>

Answer　　223,793

Steps

1. Ask a student to give you any 5-digit number and write it on the board. *23,795*

2. The teacher adds 199,998 to that number (or, to make it more simple, put the numeral 2 in front of the number the student has given you, and subtract 2 from the last digit in the number the student has given you). [223,793] The teacher writes this 6-digit number on a piece of paper, folds it over so it can't be seen, and gives it to a student to put in his or her pocket until the end of the trick.

Note: If the student gives you a 5-digit number whose last digit is 0 or 1, you will need to subtract 2 from the last two digits of his or her number. For example, if the student gives you 23,790, you would put 2 in front of that number. Then subtract 2 from 90, coming up with 223,788.

3. The teacher asks another student to give a 5-digit number and write it below the first 5-digit number given earlier. *64,028*

4. The teacher says it is time for him or her to have a turn, so the teacher writes a special 5-digit number. The digits in each of the five places (ones, tens, hundreds) of the number the second student gave the teacher plus the digits in the number used by the teacher must each total 9. In the example above, the second student gave the number 64,028, so the teacher's number must be 35,971. The 8 in the one's place of the student's number plus the 1 in the one's place of the teacher's number equals 9, the 2 in the ten's place of the student's number plus the 7 in the ten's place of the teacher's number equals 9. *35,971*

5. The teacher asks another student to give a 5-digit number. *16,264*

6. The teacher says that he or she will add a final 5-digit number, again being sure that the total of the numbers in each place is 9. *83,735*

7. The teacher then asks a student to come to the board and add up the five 5-digit numbers. (The student should get 223,793.) If the teacher sees that the student is making a mistake, then he or she can ask the student to double-check the work. *223,793*

8. The teacher then asks the student who has the folded piece of paper with the answer in his or her pocket to retrieve the piece of paper, unfold it, and read the answer aloud. Students will be amazed that the number on the piece of paper matches the answer to the addition problem.

9. The teacher then asks students to see if they can figure out how he or she knew the answer to the problem before the problem was completely written on the board.

How It Works

The key is adding 99,999 twice to the original number. Or, to look at it another way, you are adding 199,998 to the original number (99,999 × 2 = 199,998), which is just 2 short of adding 200,000. That's why—on the secret number sheet you give to a student to put in his or her pocket—you put the numeral 2 in front of the original number and subtract 2 from it. The teacher controls the problem by being sure that he or she gets a total of 9 in the ones column, the tens column, and so on, when adding the teacher's numbers to the second and fourth numbers students have given. In other words, the teacher is guaranteeing that 99,999 is added twice to the original number.

Extension

Once students have figured out the way this problem works, challenge them to see if it works with 4-digit numbers, 6-digit numbers, and so on.

Contents of the Digital Resources

Title	Filename
Paperfolding Circle	paperfolding.pdf
Questions to Elicit Mathematical Learning	questions.pdf
Before the Math Talk	beforemt.pdf
During the Math Talk	duingmt.pdf
After the Math Talk Reflection	aftermt.pdf

Notes

Notes